Department of the Army
U.S. Army Corps of Engineers

Engineering and Design

DESIGN OF PILE FOUNDATIONS

**US Army Corps
of Engineers**

ENGINEERING AND DESIGN

Design of Pile Foundations

A pile is basically a long cylinder of a strong material such as concrete that is pushed into the ground to act as a steady support for structures built on top of it. Pile foundations are used in the following situations: When there is a layer of weak soil at the surface.

ENGINEER MANUAL

DEPARTMENT OF THE ARMY
US Army Corps of Engineers

Engineering and Design
DESIGN OF PILE FOUNDATIONS

1. Purpose. This manual provides information, foundation exploration and testing procedures, load test methods, analysis techniques, allowable crite- ria, design procedures, and construction consideration for the selection, design, and installation of pile foundations. The guidance is based on the present state of the technology for pile-soil-structure-foundation interaction behavior. This manual provides design guidance intended specifically for the geotechnical and structural engineer but also provides essential information for others interested in pile foundations such as the construction engineer in understanding construction techniques related to pile behavior during instal- lation. Since the understanding of the physical causes of pile foundation behavior is actively expanding by better definition through ongoing research, prototype, model pile, and pile group testing and development of more refined analytical models, this manual is intended to provide examples and procedures of what has been proven successful. This is not the last nor final word on the state of the art for this technology. We expect, as further practical design and installation procedures are developed from the expansion of this technology, that these updates will be issued as changes to this manual.

2. Applicability. This manual is applicable to all USACE commands having civil works responsibilities, especially those geotechnical and structural engineers charged with the responsibility for design and installation of safe and economical pile foundations.

DEPARTMENT OF THE ARMY
US Army Corps of Engineers

Engineering and Design
DESIGN OF PILE FOUNDATIONS

Table of Contents

Subject	Paragraph	Page

i

CHAPTER 1

INTRODUCTION

1-1. Purpose. This manual provides information, foundation exploration and
testing procedures, load test methods, analysis techniques, design criteria
and procedures, and construction considerations for the selection, design, and
installation of pile foundations. The guidance is based on the present state
of technology for pile-soil-structure-foundation interaction behavior. This
manual provides design guidance intended specifically for geotechnical and
structural engineers and essential information for others interested in understanding construction techniques related to pile behavior during installation.
The understanding of pile foundation behavior is actively expanding by ongoing
research, prototype, model pile, and pile group testing and development of
more refined analytical models. However, this manual is intended to provide
examples and procedures of proven technology. This manual will be updated as
changes in design and installation procedures are developed.

1-2. Applicability. This manual is applicable to all USACE commands having
civil works responsibilities, especially those geotechnical and structural
engineers charged with the responsibility for design and installation of safe
and economical pile foundations.

1-3. References, Bibliographical and Related Material.

    a. US Army Corps of Engineers Directives are listed in Appendix A.

    b. Bibliographical and related material is listed in Appendix B,
numbered, and cited in the text by the corresponding item number. These
selections pertain to pile foundations for general knowledge and contain
further expanded and related material.

    c. A series of computer programs are available to assist in analyzing
and designing pile foundations in accordance with the engineering manual.
This series of programs includes:

    (1) Pile Group Analysis (CPGA) which is a stiffness analysis of three-
dimensional pile groups assuming linear elastic pile-soil interaction and a
rigid pile cap.

    (2) Pile Group Graphics (CPGG) which displays geometry and the results
of CPGA.

    (3) Pile Group Stiffness (CPGS) which determines the pile head stiffness
coefficients for a single vertical pile, and computes the displacements,
internal forces and moments, and axial and lateral soil pressures acting on a
pile due to specified loads or displacements at the pile head.

    (4) Pile Group Dynamics (CPGD) which extends the capability of CPGA to
account for dynamic loading.

    (5) Pile Group Concrete (CPGC) which develops the interaction diagrams
and data required to investigate the structural capacity of prestressed
concrete piles.

(6) Pile Group Interference (CPGI) which investigates the pile layout for geometric interference due to the intersection of piles during driving.

(7) Pile Group Optimization (CPGO) which solves for the optimal arrangement of a pile group using data and analysis results from GPGA.

(8) Pile Group Base (CPGB) which analyzes a rigid base slab or pile cap for pile loads determined by CPGA.

(9) Pile Group Flexible (CPGF) which extends the capability of CPGA to account for the flexibility of the base slab or pile cap.

(10) Pile Group Probabilistic (CPGP) which extends the capability of CPGI to account for probabilistic variations in pile driving tolerances, tolerable manufacturing imperfections, pile flexibility, and subsurface obstructions.

The first five programs are available for use, and the remaining programs are under development. Other programs will be added to the series as needs are identified. Currently available programs are fully described in Items 5, 6, 15, and 16, respectively. The theoretical background for these computer programs and this Engineer Manual will be provided in "Theoretical Manual for the Design of Pile Foundations." The Theoretical Manual is currently in preparation and is intended to be a companion volume that provides a detailed discussion of the techniques used for the design/analysis of pile foundations as presented in this manual and used in the available computer programs listed on pp 1-1 and 1-2. It will present the theoretical development of these engineering procedures, how they were implemented in computer programs, and discussions on the limitations of each method.

d. A case history of pile driving at Lock and Dam No. 1, Red River Waterway, is presented in Appendix C.

e. Examples of pile capacity computations are presented in Appendix D.

1-4. Definitions.

a. Pile Foundation. In this manual, a pile foundation will be broadly described as one in which the following is true of the piles:

(1) Piles are driven, not drilled.

(2) Standard commercial, not special patent, piles are used.

(3) Usually steel or prestressed concrete piles are used for major hydraulic structures, but reinforced concrete or timber piles should also be considered.

b. Pile Industry Terms. Since many of the terms used in the piling (material, equipment, and driving) industry seem to be unique to this industry, it is suggested that reference be made to the Deep Foundations Institute (Item 32). These definitions are adhered to in this manual.

c. Units of Measurement. The English system of measurement units have been used exclusively throughout this manual.

d. Notations and Symbols. There is no unified set of symbols and nomenclature for the analysis and design of pile groups. Pile technology has
evolved over the last three decades and different symbols appear throughout
the engineering literature for describing the various geotechnical and structural aspects of single pile and pile group behavior. This has presented a
major problem in writing this EM. The following approach was adopted:

(1) It was not practical to develop a unified system of symbols and
nomenclature.

(2) Critical symbols which have attained recognition as defacto standards throughout the profession (such as p-y and t-z curves) and within the
Corps of Engineers (such as X, Y, and Z for the global coordinate axes and 1,
2, and 3 for the local pile coordinate axes) will be identified. Some symbols
may therefore have dual meanings (such as x, y, and z for local coordinates or
as local pile displacements).

e. Style of Presentation. The EM was written in a manner to assist
readers struggling with the difficulties of the symbols and nomenclature and
the inherent technical complexity of pile behavior. Footnotes are used when a
symbol which has a dual meaning is introduced. This minimizes potential problems by explaining the meaning for that particular application and gives the
key references for a detailed explanation.

f. Alternative Foundation Designs. The first consideration in the design of a structural foundation should be the subsurface investigation. The
data from such investigations should be evaluated to determine whether or not
the use of a pile foundation is necessary. If such studies, together with
studies of the soil properties, reveal that detrimental settlement can be
avoided by more economical methods or that a pile foundation will not prevent
detrimental settlement, then piles should not be used. A preliminary selection of the pile type may usually be made from a study of the foundation
investigations. However, the nature of the structure, type of applied loads,
and technical and economic feasibility must be considered in the determination
of pile type, length, spacing, batters, etc.

(1) If the boring data reveal that timber piles would not be damaged by
driving, such type may be considered. Steel bearing piles may be desirable if
boulders or hard strata are present in the area of pile driving. In deposits
of sands, silts, and clays that are relatively free of boulders, consideration
should be given to the use of concrete piles. However, considerable difficulty and problems often occur in driving displacement piles in granular soils
such as sands, silty-sands, and sandy silts.

(2) The load-bearing stratum or strata can be selected from a study of
the soil profiles and characteristics of the soil. By estimating the required
length of penetration into the load-bearing material, the lengths of piles may
be reasonably approximated. In designing friction pile foundations, advantage
should be taken of increased capacity with greater depths by favoring fewer
piles with greater lengths.

(3) In cases where piles are to be driven into or underlain by cohesive
soils, the foundation should be investigated to determine the type and length
of piles and the size and shape of the structural foundation which will result
in a minimum of ultimate settlement. In wide foundations, long, heavily

loaded, widely spaced piles will result in less settlement than short, lightly loaded, closely spaced piles.

CHAPTER 2

GENERAL CONSIDERATIONS

2-1. General. Many factors must be considered when selecting an appropriate
foundation for a hydraulic structure. This chapter presents criteria and
methods for selecting the best type of foundation. Information is presented
to identify the feasible foundation alternatives for more detailed study. The
final selection should be based on an evaluation of engineering feasibility
and comparative costs for the potential alternatives considering such factors
as safety, reliability, constructability, and life cycle performance. This
chapter also presents general criteria for feature design. Such criteria
pertain to the type and function of the structure, the nature of the applied
loads, and the type of foundation material. The requirements for a subsurface
investigation program are also presented.

2-2. Structural and Geotechnical Coordination. A fully coordinated effort
from geotechnical and structural engineers and geologists should ensure that
the result of the pile foundation analysis is properly integrated into the
overall foundation design. This coordination extends through plans and
specifications, preconstruction meetings, and construction. Some of the
critical aspects of the design process which require coordination are:

    a. Preliminary and final selection of pile type.

    b. Allowable deflections at the groundline and fixity of the pile head.

    c. Preliminary evaluation of geotechnical data and subsurface
conditions.

    d. Selection of loading conditions, loading effects, potential failure
mechanisms, and other related features of the analytical models.

    e. Minimum pile spacing and maximum batter.

    f. Lateral resistance of soil.

    g. Required pile length and axial capacity.

    (1) Maximum stresses during handling, driving, and service loading.

    (2) Load testing and monitoring programs.

    h. Driveability of the pile to the selected capacity.

2-3. Design Considerations. The pile foundation analysis is based upon
several simplifying assumptions which affect the accuracy of the results. The
computed results must always be reviewed with engineering judgement by the
design engineer to assure that the values are reasonable. Also, the analysis
results should be compared with load test results.

    a. Functional Significance of Structure. The type, purpose, and func-
tion of the structure affect decisions regarding subsurface investigation programs, analytical methods, construction procedures and inspection, and
performance monitoring. Generally, the proposed structure should be evaluated

on the basis of the consequences of failure, that is, the potential for loss of lives and property, economic losses both local and national, compromising the national defense, and adverse public opinion. The designer must be aware of these factors so that a rational approach may be taken throughout the analysis, design, and construction of the project. In order to reduce the potential for failure, as well as to minimize the cost, the designer must apply appropriate factors of safety to the design. These factors of safety are based on the functional significance of the structure, the level of confidence in the foundation parameters, the adequacy of the analysis tools, and the level of construction controls.

b. Definitions of Failure. Structure or foundation failures can be categorized as an actual collapse or a functional failure. Functional failure can be due to excessive deflection, unacceptable differential movements, excessive vibration, and premature deterioration due to environmental factors. For critical structures, failure to meet functional requirements may be as serious as the actual collapse of a lesser structure. Therefore, designers should be cognizant not only of the degree of safety against collapse but also of effects of settlement and vibration on the functional performance.

c. Factors of Safety. Factors of safety represent reserve capacity which a foundation or structure has against collapse for a given set of loads and design conditions. Uncertain design parameters and loads, require a higher factor of safety than required when the design parameters are well known. For most hydraulic structures, designers should have a high level of confidence in the soil and pile parameters and the analysis. Therefore, uncertainty in the analysis and design parameters should be minimized rather than requiring a high factor of safety. For less significant structures, it is permissible to use larger factors of safety if it is not economical to reduce the uncertainty in the analysis and design by performing additional studies, testing, etc. Also, factors of safety must be selected to assure satisfactory performance for service conditions. Failure of critical components to perform as expected can be as detrimental as an actual collapse. Therefore, it is imperative that in choosing a design approach, the designer consider the functional significance of the project, the degree of uncertainty in the design parameters and the analytical approach, and the probability of failure due to both collapse and functional inadequacy.

d. Soil-Structure Considerations for Analysis. The functional significance and economic considerations of the structure will determine the type and degree of the foundation exploration and testing program, the pile test program, the settlement and seepage analyses, and the analytical models for the pile and structure. For critical structures the foundation testing program should clearly define the necessary parameters for the design of the pile foundation, such as soil types and profiles, soil strengths, etc. (Paragraphs 3-1 and 3-2 give further details.) Although pile load tests are usually expensive and time consuming, they are invaluable for confirming or modifying a pile foundation design during the construction phase. A well-planned and monitored pile load test program will usually save money by allowing the designer to utilize a lower factor of safety or by modifying the required number or length of piles required. A pile load test program should be considered for all large structures for which a pile foundation is required. (Paragraph 3-6 gives further details.) Depending upon the type of foundation material, the nature of the loading, the location of the ground water, and the functional requirements of the structure, a detailed seepage

analysis and/or pile settlement analysis may also be required to define
adequately the pile-soil load transfer mechanism and the resulting parameters
necessary for an adequate pile design. Where differential movement between
monoliths is a concern, an accurate estimate of pile settlement may be
crucial, particularly if the monoliths have significantly different load
levels. (Paragraphs 3-4 and 4-4 give further discussions.) Decisions
regarding the type and sophistication of the analytical models for the pile
and the structure should also be made with the functional significance of the
structure in mind. For example, it may be satisfactory to analyze the pile
foundation for a small, lightly loaded structure based on conservative
assumptions for pile parameters and a crude structural model; however, a
larger, more important structure would probably require a detailed single pile
analysis to establish the proper pile parameters. Perhaps it would even be
necessary to use a structural model capable of considering the actual structural stiffness to insure correct load distribution to the piles. (See paragraph 4-5 for further discussion.)

    e. Construction and Service Considerations. No matter how thorough and
well researched a design may be, it is only as good as its execution in the
field. The proof of the entire design and construction process is in the
performance of the final product under service conditions. Therefore, the
designer should consider the analysis and design of a structure and its
foundation as parts of an engineering process that culminates with the
successful long-term performance of the structure for its intended purposes.
The designer prepares the specifications and instructions for field personnel
to assure the proper execution of the design. The designer must discuss critical aspects of the design with construction personnel to make sure that there
is a thorough understanding of important design features. For critical
structures a representative of the design office should be present in the
field on a continuous basis. One such example would be a major pile test
program where the execution of the pile test and the gathering of data is
critical for both a successful testing program and verification of design
assumptions. Another critical activity that requires close cooperation
between the field and the designer is the installation of the foundation
piling. The designer should be involved in this phase to the extent necessary
to be confident that the design is being properly executed in the field. As a
general principle, designers should make frequent visits to the construction
site not only to ensure that the design intent is being fulfilled but also to
familiarize themselves with construction procedures and problems to improve on
future designs and complete as-built records. Once the project is in operation, the designer should obtain feedback on how well the structure is
fulfilling its operational purposes. This may require that instrumentation be
a part of the design or may take the form of feedback from operating personnel
and periodic inspections.

2-4. Nature of Loadings.

    a. Usual. Usual loads refer to conditions which are related to the pri-
mary function of a structure and can be reasonably expected to occur during
the economic service life. The loading effects may be of either a long term,
constant or an intermittent, repetitive nature. Pile allowable loads and
stresses should include a conservative safety factor for such conditions. The
pile foundation layout should be designed to be most efficient for these
loads.

b. Unusual. Unusual loads refer to construction, operation or maintenance conditions which are of relatively short duration or infrequent occurrence. Risks associated with injuries or property losses can be reliably
controlled by specifying the sequence or duration of activities, and/or by
monitoring performance. Only minor cosmetic damage to the structure may occur
during these conditions. Lower factors of safety may be used for such loadings, or overstress factors may be applied to the allowables for these loads.
A less efficient pile layout is acceptable for these conditions.

c. Extreme. Extreme loads refer to events which are highly improbable
and can be regarded as emergency conditions. Such events may be associated
with major accidents involving impacts or explosions and natural disasters due
to earthquakes or hurricanes which have a frequency of occurrence that greatly
exceeds the economic service life of the structure. Extreme loadings may also
result from a combination of unusual loading effects. The basic design concept for normal loading conditions should be efficiently adapted to accommodate extreme loading effects without experiencing a catastrophic failure.
Extreme loadings may cause significant structural damage which partially
impairs the operational functions and requires major rehabilitation or replacement of the structure. The behavior of pile foundations during extreme
seismic events is a phenomenon which is not fully understood at present. The
existing general approach is to investigate the effects of earthquake loading
at sites in seismic Zones 1 or 2 by applying psuedostatic forces to the
structure and using appropriate subgrade parameters. In Zones 3 or 4 a
dynamic analysis of the pile group is appropriate. Selection of minimum
safety factors for extreme seismic events must be consistent with the seismol-
ogic technique used to estimate the earthquake magnitude. Designing for pile
ductility in high risk seismic regions is very important because it is very
difficult to assess pile damage after earthquakes and the potential repair
costs are very large. Effects related to liquefaction of subsurface strata
are discussed in paragraph 3-5.

2-5. Foundation Material.

a. Known Data. After a general site for a project is selected, the designer should make a site visit to examine the topography at the site. Rock
outcrops or highway cuts on or near the site may provide valuable information
of the subsurface conditions. An examination of existing structures in the
vicinity may also provide information. A visit to the local building department may provide foundation information and boring logs for nearby buildings.
The highway department may have soil and geological information in the area
for existing roads and bridges. Valuable soil and geological information can
be obtained from other governmental agencies, such as the United States
Geological Survey (USGS), Soil Conservation Service (SCS), Bureau of Records,
etc., for even remotely located areas. Colleagues may be able to provide
information on projects they have worked on in the area. Check the files for
previous jobs your office might have built or explored in the area.

b. Similar Sites. It is important to determine the geological history
of the site and geological origins of the material that exists at the site.
The geological history of the site will provide information on the properties
of the different geological zones and may allow the designer to find sites
with similar geological origins where data are available on the soil and rock
properties and on pile behavior.

c. Exploration Requirements. The designer must lay out an exploration and testing program that will identify the various material zones and their properties. This exploration and testing program shall identify the various soil and rock layers at the site; the groundwater table, water quality, and existing aquifers; and information relating to faults at the site. The above information should be obtained to the degree that is necessary to design an adequate foundation for the proposed structure.

2-6. Identification and Evaluation of Pile Alternatives.

a. General. Structures may be founded on rock, on strong or weak soils, cohesive or noncohesive soils, above ground level, below water level, etc. The type of foundation used to support a structure depends on local conditions. After obtaining a general evaluation of the subsurface conditions the engineer should attempt to identify all potential useful foundation alternatives for a structure. Three basic types of foundations are available: soil-founded, various types of piles, and piers or caissons. Each of these foundation types has many subcategories. The following paragraphs provide a short description and evaluation of the various pile types.

b. Piles. The purpose of a pile foundation is to transfer and distribute load through a material or stratum with inadequate bearing, sliding or uplift capacity to a firmer stratum that is capable of supporting the load without detrimental displacement. A wide range of pile types is available for applications with various soil types and structural requirements. A short description of features of common types of piles follows:

(1) Steel H-Piles. Steel H-piles have significant advantages over other types of piles. They can provide high axial working capacity, exceeding 400 kips. They may be obtained in a wide variety of sizes and lengths and may be easily handled, spliced, and cut off. H-piles displace little soil and are fairly easy to drive. They can penetrate obstacles better than most piles, with less damage to the pile from the obstacle or from hard driving. The major disadvantages of steel H-piles are the high material costs for steel and possible long delivery time for mill orders. H-piles may also be subject to excessive corrosion in certain environments unless preventive measures are used. Pile shoes are required when driving in dense sand strata, gravel strata, cobble-boulder zones, and when driving piles to refusal on a hard layer of bedrock.

(2) Steel Pipe Piles. Steel pipe piles may be driven open- or closed-end and may be filled with concrete or left unfilled. Concrete filled pipe piles may provide very high load capacity, over 1,000 kips in some cases. Installation of pipe piles is more difficult than H-piles because closed-end piles displace more soil, and open-ended pipe piles tend to form a soil plug at the bottom and act like a closed-end pile. Handling, splicing, and cutting are easy. Pipe piles have disadvantages similar to H-piles (i.e., high steel costs, long delivery time, and potential corrosion problems).

(3) Precast Concrete. Precast concrete piles are usually prestressed to withstand driving and handling stresses. Axial load capacity may reach 500 kips or more. They have high load capacity as friction piles in sand or where tip bearing on soil is important. Concrete piles are usually durable and corrosion resistant and are often used where the pile must extend above ground. However, in some salt water applications durability is also a problem

with precast concrete piles. Handling of long piles and driving of precast concrete piles are more difficult than for steel piles. For prestressed piles, when the required length is not known precisely, cutting is much more critical, and splicing is more difficult when needed to transfer tensile and lateral forces from the pile head to the base slab.

(4) Cast-in-Place Concrete. Cast-in-place concrete piles are shafts of concrete cast in thin shell pipes, top driven in the soil, and usually closed end. Such piles can provide up to a 200-kip capacity. The chief advantage over precast piles is the ease of changing lengths by cutting or splicing the shell. The material cost of cast-in-place piles is relatively low. They are not feasible when driving through hard soils or rock.

(5) Mandrel-Driven Piles. Mandrel-driven piles are thin steel shells driven in the ground with a mandrel and then filled with concrete. Such piles can provide up to a 200-kip capacity. The disadvantages are that such piles usually require patented, franchised systems for installation and installation is not as simple as for steel or precast concrete piles. They offer the advantage of lesser steel costs since thinner material can be used than is the case for top-driven piles. The heavy mandrel makes high capacities possible. Mandrel-driven piles may be very difficult to increase in length since the maximum pile length that can be driven is limited by the length of the mandrel available at the site. Contractors may claim extra costs if required to bring a longer mandrel to the site.

(6) Timber. Timber piles are relatively inexpensive, short, low-capacity piles. Long Douglas Fir piles are available but they will be more expensive. They may be desirable in some applications such as particular types of corrosive groundwater. Loads are usually limited to 70 kips. The piles are very convenient for handling. Untreated timber piles are highly susceptible to decay, insects, and borers in certain environments. They are easily damaged during hard driving and are inconvenient to splice.

c. Evaluation of Pile Types.

(1) Load Capacity and Pile Spacing. Of prime importance is the load-carrying capacity of the piles. In determining the capacity of a pile foundation, it is important to consider the pile spacing along with the capacity of individual piles. The lateral load resistance of the piles may also be important since lateral loads can induce high bending stresses in a pile.

(2) Constructability. The influence of anticipated subsurface and surface effects on constructability must be considered. Piles susceptible to damage during hard driving are less likely to penetrate hard strata or gravel and boulder zones. Soil disturbance or transmission of driving vibrations during construction may damage adjacent piles or structures. Pile spacing and batters must be selected to prevent interference with other structural components during driving. The ease of cutting or splicing a pile may also affect constructability.

(3) Performance. The pile foundation must perform as designed for the life of the structure. Performance can be described in terms of structural displacements which may be just as harmful to a structure as an actual pile failure. The load capacity should not degrade over time due to deterioration of the pile material.

(4) Availability. Piles must be available in the lengths required, or they must be spliced or cut off. Project scheduling may make lead time an important consideration, since some piles may require up to 6 months between order and delivery.

(5) Cost. Once a pile type satisfies all other criteria, relative cost becomes a major consideration. For comparisons between types of piles, it may be adequate to compare the pile cost per load capacity. For example, an installed H-pile may cost $40 per linear foot and have a capacity of 200 kips for a 50-foot length. The unit capacity cost would then be $10 per kip. A comparison between unit capacity costs may lead to an obvious exclusion of certain pile types. The cost evaluation should include all expenses related to and dependent on the pile foundation. Such costs may include additional expense for storage or splicing. They may include pressure-relief systems used to reduce uplift pressures and thus control pile loads. In addition, any required modifications to the structure to accommodate the piles should be included in a comparative cost estimate. For example, an increase in base slab thickness may be required to provide additional embedment for the tops of the piles.

d. Preliminary Evaluations. All identified foundation alternatives should first be evaluated for suitability for the intended application and cost. For piles, this evaluation should be based on the capacity, availability, constructability, and expected performance of the various types of piles. Initial evaluation of nonpile alternatives should be based on similar criteria. This will limit further studies to those foundation alternatives which are reasonably feasible. During this initial evaluation, it may also be possible to eliminate from consideration obvious high-cost alternatives.

e. Final Evaluations. The final evaluation and selection should be based mainly on relative costs of the remaining alternatives. This evaluation should include the costs of structural or site modifications required to accommodate the foundation type. Cost and other factors may be important in the selection. Differences in delivery or installation schedules, levels of reliability of performance, and potential construction complications may be considered. When comparing a pile foundation to another type of foundation, it will be necessary to develop a preliminary pile layout to determine a reasonable estimate of quantities.

2-7. Field Responsibilities for the Design Engineer.

a. Loading Test. On all major structures with significant foundation costs, pile load tests are required. On minor structures, pile load tests may not be required depending on economics, the complexity of the soil conditions, the loading conditions and the experience the designer has with pile foundations in that area. Load tests of piles should be performed to finalize pile lengths and to provide information for improving design procedures. Load tests are performed prior to construction of the pile foundation. Consideration should be given to the use of indicator pile tests, that is the capacity may be inferred using the pile driving analyzer or other similar technique. These are powerful tools that can augment but not replace static tests.

b. Field Visits. The quality design of a pile foundation design is only as good as the as-built conditions. In order to ensure correct installation of the pile foundation, it is important for the design engineer to visit the construction site frequently. Field visits should be made to view critical construction phases and to discuss progress and potential changes in procedures with the construction representative. Critical items include monitoring and maintaining detailed records of driving operations, especially:

   (1) Driving reports for individual piles - date and times, placement position and alinement; blow counts, difficulties and interruptions during driving; installation and location of any pile splices.

   (2) General driving data - complete descriptions of driving equipment, adjustments and changes (leads, hammer, anvil, cap, cushions, etc.); pile storage and handling procedures; pile interference; pile heave.

   (3) Driving restrictions - existing structures in vicinity; driving near new concrete; limiting water elevation.

   c. Instructions to the Field. Instructions to the field are necessary to convey to field personnel the intent of the design. Instructions to the field should be conveyed to the field by the designers through a report, "Engineering Considerations and Instructions for Field Personnel" as required by EM 1110-2-1910. This report should include the following information to the field:

   (1) Present the design assumptions made regarding interpretation of subsurface investigation data and field conditions.

   (2) The concepts, assumptions, and special details of the design.

   (3) Assistance to field personnel in interpreting the plans and specifications.

   (4) Information to make field personnel aware of critical areas in the design which require additional control and inspection.

   (5) Provide results of wave equation analysis with explanation of application of results to monitor driving operations.

   (6) Provide guidance for use of pile driving analyzer to monitor driving operations.

2-8. Subsurface Conditions. The ultimate axial load capacity of a single pile is generally accepted to be the total skin friction force between the soil and the pile plus the tip capacity of the pile, which are dependent on the subsurface conditions. The ultimate axial capacity of individual friction piles depends primarily upon the type of soil: soft clay, stiff clay, sand, or stratified soil layers. In soil deposits that contain layers of varying stiffness, the ultimate axial pile capacity cannot be equal to the sum of the peak strength of all the materials in contact with the pile because the peak strengths are not reached simultaneously. Failure is likely to be progressive. The existence of boulders or cobbles within foundation layers can present driving problems and hinder determination of ultimate axial capacity of a single pile.

2-9. Pile Instrumentation. Pile instrumentation can be delineated into three categories: instrumentation used during pile load tests to obtain design data, pile driving analyzer used to control quality of pile installation, and permanent instrumentation used to gather information during the service life of the project. Decisions on the type of instrumentation for pile load tests must be an integral part of the design. The designer should select instrumentation that has sufficient accuracy to measure the required data. Permanent instrumentation is used to gather data relating to the state of stress and behavior of the pile under service load conditions. Useful knowledge can be gained from permanent instrumentation, not only about the behavior of a particular pile foundation, but also about analysis and design assumptions in general. Verification (or modification) can be obtained for analytically derived information such as pile settlement, pile head fixity, location of maximum moment within the pile, and the distribution of loads to an individual pile within a group. However, a permanent instrumentation program can be very expensive and should be considered only on critical projects. Also, effective use of the instrumentation program depends on a continuing commitment to gather, reduce, and evaluate the data.

CHAPTER 3

GEOTECHNICAL CONSIDERATIONS

3-1. Subsurface Investigations and Geology. The subsurface explorations are
the first consideration from site selection through design. These investigations should be planned to gain full and accurate information beneath and
immediately adjacent to the structure. The investigation program should cover
the area of the foundation and, as a very minimum, extend 20 feet below the
tip of the longest pile anticipated. The borings should be of sufficient
depth below the pile tip to identify any soft, settlement-prone layers. The
type of soil-boring will be determined by the type of soil profile that
exists. In a clay layer or profile, sufficient undisturbed samples should be
obtained to determine the shear strength and consolidation characteristics of
the clay. The sensitivity of the clay soils will have to be determined, as
strength loss from remolding during installation may reduce ultimate pile
capacity. Shrink-swell characteristics should be investigated in expansive
soils, as they affect both capacity and movement of the foundation. Since
most structures requiring a pile foundation require excavation that changes
the in situ soil confining pressure and possibly affects the blow count, the
standard penetration test commonly performed in granular soils will probably
be of limited use unless the appropriate corrections are made. It should be
understood, however, that the standard penetration test is valid when applied
properly. Where gravels or cobbles are expected, some large diameter soil
borings should be made in order to collect representative samples upon which
to determine their properties. An accurate location of the soil borings
should be made in the field and a map provided in the design documents. An
engineering geologist should be present during the drilling operation to
provide field interpretation. The geologist must have the latitude to re-
locate the borings to define the subsurface profile in the best way possible.
Geologic interpretations should be provided in the design documents in the
form of geologic maps and/or profiles. The profiles should extend from the
ground surface to well below the deepest foundation element. The accompanying
text and/or maps should fully explain the stratigraphy of the subgrade as well
as its engineering geology characteristics.

3-2. Laboratory and Field Testing. Laboratory determinations of the shear
strength and consolidation properties of the clay and clayey soils should be
performed routinely. For details of performing the individual test, refer to
the laboratory test manual, EM-1110-2-1906. For the construction case in clay
soils, the unconsolidated-undrained triaxial shear (Q) test should be performed. In silts, the consolidated-undrained triaxial shear (R) test, with
pore pressure recorded, should be performed and used to predict the shear
strength of the formation appropriate to the construction and long-term loading cases. In sands, the standard penetration test, or if samples can be
collected, the consolidated-drained triaxial shear test or direct shear test
(S) should be used to predict the shear strength appropriate to the two loading cases. The sensitivity of these soils should be estimated and the appropriate remolded triaxial shear test performed, as well as the shrink-swell
tests, if appropriate. Consolidation tests should be performed throughout the
profile so that the downdrag and/or settlement of the structure may be
estimated. The field testing should include in situ ground-water evaluation.
In situ testing for soil properties may also be used to augment the soil borings but should never be used as a replacement. Some of the more common methods would be the electronic cone penetration test, vane shear, Swedish vane

borer, or pressuremeter. Geophysical techniques of logging the soil boring, electric logging, should be employed wherever possible if highly stratified soils are encountered or expected or if faults need to be located.

3-3. Foundation Modification. Installation of piles will densify loose, granular materials and may loosen dense, granular materials. This should be taken into consideration by the designer. For homogeneous stratifications, the best pile foundations would tend theoretically toward longer piles at a larger spacing; however, if the piles densify the granular soils, pile driving may become impossible at great depth. Pile installation affects soils from about 5 feet to 8 pile tip diameters laterally away from the pile and vertically below the tip; therefore, the designer should exercise judgement as to the effect that driving will have upon the foundation. In silty subgrades, the foundations may dilate and lose strength which will not be regained. Piles can be used to modify foundation soils by densification, but pile driving may be a costly alternative to subgrade vibration by other means. In soft clay soils piles could be used to achieve some slight gain in shear strength; however, there are more cost effective methods to produce the same or better results, such as surcharge and drainage. It may be necessary to treat piles or soil to provide isolation from consolidation, downdrag, or swell. This treatment may be in the form of prebored larger diameter cased holes or a material applied to the pile to reduce adhesion.

3-4. Groundwater Studies. The groundwater should be evaluated in each of the soil borings during the field investigation. Piezometers and/or monitoring wells should be installed and monitored during the various weather cycles. A determination should be made of all of the groundwater environments beneath the structure, i.e. perched water tables, artesian conditions and deep aquifers. The field tests mentioned in paragraph 3-2 will be useful in evaluating the movement of groundwater. Artesian conditions or cases of excess pore water pressure should also be considered as they tend to reduce the load-carrying capacity of the soil. An effective weight analysis is the best method of computing the capacity of piles. For the design of pile foundations the highest groundwater table elevation should prove to be the worst case for analysis of pile capacity. However, significant lowering of the water table during construction may cause installation and later service problems by inducing densification or consolidation.

3-5. Dynamic Considerations. Under dynamic loading, radical movements of the foundation and/or surrounding area may be experienced for soils that are subject to liquefaction. Liquefaction is most commonly induced by seismic loading and rarely by vibrations due to pile driving during construction or from vibrations occurring during operations. For dynamic loadings from construction or operations, the attenuation of the vibrations through the foundation and potential for liquefaction should be evaluated. In seismic Zones 2, 3, and 4, the potential liquefaction should be evaluated for the foundations. If soils in the foundation or surrounding area are subject to liquefaction, the removal or densifaction of the liquefiable material should be considered, along with alternative foundation designs. The first few natural frequencies of the structure-foundation system should be evaluated and compared to the operating frequencies to assure that resonance (not associated with liquefaction) is not induced.

3-6. Pile Load Test.

    a. General. The pile load test is intended to validate the computed
capacity for a pile foundation and also to provide information for the
improvement of design rational. Therefore, a test to pile failure or soil/
pile failure should be conducted in lieu of testing to a specified load of
termination. Data from a test should not be used to lengthen or shorten piles
to an extent that their new capacities will vary more than 10 percent from the
test load. Finally, if the pile tests are used to project pile capacity for
tip elevations other than those tested, caution should be exercised. In a
complex or layered foundation, selecting a tip elevation for the service piles
different from the test piles may possibly change the pile capacity to values
other than those projected by the test. As an example, shortening the service
piles may place the tips above a firm bearing stratum into a soft clay layer.
In addition to a loss in bearing capacity, this clay layer may consolidate
over time and cause a transfer of the pile load to another stratum. Lengthening the service piles may cause similar problems and actually reduce the load
capacity of the service piles if the tips are placed below a firm bearing
stratum. Also, extending tips deeper into a firmer bearing may cause driving
problems requiring the use of jetting, predrilling, etc. These techniques
could significantly alter the load capacity of the service piles relative to
the values revealed by the test pile program. A pile load testing program
ideally begins with the driving of probe piles (piles driven at selected
locations with a primary intention of gaining driving information) to gain
knowledge regarding installation, concentrating their location in any suspect
or highly variable areas of the foundation strata. Test piles are selected
from among the probe piles based upon evaluation of the driving information.
The probe and test piles should be driven and tested in advance of the
construction contract to allow hammer selection testing and to allow final
selection of the pile length. Upon completion of the testing program, the
probe/test piles should be extracted and inspected. The test piles, selected
from among the probe piles driven, should be those driven with the hammer
selected for production pile driving if at all possible. In some cases
different hammers will produce piles of different ultimate capacity. Additionally, use of the production hammer will allow a correlation between blow
count and pile capacity which will be helpful during production pile driving.
The pile driving analyzer should be used wherever possible in conjunction with
the probe/test piles. This will allow the pile driving analyzer results to be
correlated with the static tests, and greater reliance can be placed upon
future results when using the analyzer for verifying the driving system
efficiency, capacity, and pile integrity for production piles.

    b. Safety Factor for Design. It is normal to apply safety factors to
the ultimate load predicted, theoretically or from field load tests. These
safety factors should be selected judiciously, depending upon a number of
factors, including the consequences of failure and the amount of knowledge
designers have gained relative to the subsurface conditions, loading conditions, life of the structure, etc. In general, safety factors for hydraulic
structures are described in paragraph 4-2C.

    c. Basis for Tests. A pile loading test is warranted if a sufficient
number of production piles are to be driven and if a reduced factor of safety
(increased allowable capacity) will result in a sufficient shortening of the
piles so that a potential net cost savings will result. This is based upon
the assumption that when a test pile is not used, a higher safety factor is

required than when test piles are used. If very few piles are required,
longer piles as required by the higher factor of safety (3.0) may be less
expensive than performing a pile load test, reducing the factor of safety to
2.0, and using shorter piles. Pile load tests should also be performed if the
structure will be subjected to very high loads, cyclic loads of an unusual
nature, or where highly variable soil conditions exist. Special pile load
tests should be performed to determine soil parameters used in design when the
structure is subject to large dynamic loads, such as large reciprocating
machinery, earthquakes, etc.

     d. Test Location. The pile load test should be conducted near the base
of the structure with the excavation as nearly complete as possible. If the
pile load test cannot be performed with the excavation completed, it will be
necessary to evaluate and compensate for the additional soil confining
pressure that existed during the load test. Note that casing off soils that
will later be excavated does not provide a solution to this problem. Test
piles should be located so that they can be incorporated into the final work
as service piles if practical.

     e. Cautions. A poorly performed pile load test may be worse than having
no test at all. All phases of testing and data collection should be monitored
by an engineer familiar with the project and pile load test procedures and
interpretation. In highly stratified soils where some pile-tip capacity is
used in design computations, care should be taken to keep at least 5 feet or
8 pile tip diameters of embedment into the bearing stratum. Similarly, the
tip should be seated a minimum of 5 feet or 8 pile tip diameters above the
bottom of the bearing stratum. The driving records of any piles driven should
be used to evaluate driveability of the production piles, considering the
possibility of soil densification. In clay formations, where the piles may
tend to creep under load, add in holding periods for the load test and make
sure that the load on the pile is held constant during the holding period. A
reduction in allowable load may be necessary due to settlement under long-term
sustained load (creep). The jack and reference beam should be in the same
plane with the axis of the test pile since deviations will result in erroneous
pile load tests.

3-7. Selection of Shear Strength Parameters. Based upon the geologic interpretation of the stratification, similar soil types may be grouped together
for purposes of analysis. From the triaxial shear test and any other indicator type testing, a plot of both undrained shear strength and soil unit weight
should be plotted versus depth below ground surface. If the data appear
similar in this type of display, then an average trend of undrained shear
strength and soil unit weight may be selected to typify the subgrade clays and
clayey soils. The same procedures would be followed for silty soils with the
exception that the undrained shear strength would be determined from
consolidated-undrained triaxial shear tests (R) with pore pressure measurements. This would be a construction case or short-term loading case, as the
Q case is called. For the long-term case, the shear testing would be represented by the consolidated-drained triaxial shear test or direct shear test
(S) in all soil types. The cases referenced above are shear strength cases of
the soil based upon the soil drainage conditions under which the structural
loadings will be applied. The construction case is the rapid loading without
pore pressure dissipation in the clay or clayey and silty soils represented by
the Q case. The long-term case allows drainage of the soils before or during
loading which is in general represented by the S test. This does not imply

that the construction case should not include all loads upon the completed structure. Using the shear strength data from the S test, a soil strength profile may be developed using the following equation

$$s = (\Sigma h_i \gamma_i') \tan \phi + c \qquad (3\text{-}1)$$

where

s = shear strength of the soil

h_i = height of any stratum i overlying the point at which the strength is desired

γ_i' = effective unit weight in any stratum i above the point at which the strength is desired

ϕ = angle of internal friction of the soil at the point at which the strength is desired

c = cohesion intercept of the soil at the point at which the strength is desired

The two allowable pile capacities obtained for undrained and drained soil conditions should be compared and the lower of the two cases selected for use for any tip penetration. When the design is verified by pile load test, the pile load test will take precedence in the selection of ultimate pile capacity and pile tip over the predicted theoretical value in most cases. However, the test methodology should be compatible with the predicted failure mode; that is if in the predictions the S case shear strength governs, then a Quick Test should not be selected since it will best emulate the Q case. In cases where the S case governs, then the classic slow pile test should be selected. The designer should also consider using 24-hour load holding periods at 100, 200, and 300 percent of design load especially when foundation soils are known to exhibit a tendency to creep. The load test should also include rebound and reload increments as specified in the American Society for Testing and Materials (ASTM) procedures. The uses of these shear strength parameters are explained in Chapter 4.

CHAPTER 4

ANALYSIS AND DESIGN

4-1. <u>General</u>. Design of a pile foundation involves solving the complex problem of transferring loads from the structure through the piles to the under-
lying soil. It involves the analysis of a structure-pile system, the analysis
of a soil-pile system, and the interaction of the two systems, which is highly
nonlinear. Close cooperation between the structural engineers and geotechnical engineers is essential to the development of an effective design. This
chapter addresses the criteria, procedures, and parameters necessary for the
analysis and design of pile foundations.

4-2. <u>Design Criteria</u>.

   a. Applicability and Deviations. The design criteria set forth in this
paragraph are applicable to the design and analysis of a broad range of piles,
soils and structures. Conditions that are site-specific may necessitate variations which must be substantiated by extensive studies and testing of both
the structural properties of the piling and the geotechnical properties of the
foundation.

   b. Loading Conditions.

   (1) Usual. These conditions include normal operating and frequent flood
conditions. Basic allowable stresses and safety factors should be used for
this type of loading condition.

   (2) Unusual. Higher allowable stresses and lower safety factors may be
used for unusual loading conditions such as maintenance, infrequent floods,
barge impact, construction, or hurricanes. For these conditions allowable
stresses may be increased up to 33 percent. Lower safety factors for pile
capacity may be used, as described in paragraph 4-2c.

   (3) Extreme. High allowable stresses and low safety factors are used
for extreme loading conditions such as accidental or natural disasters that
have a very remote probability of occurrence and that involve emergency
maintenance conditions after such disasters. For these conditions allowable
stresses may be increased up to 75 percent. Low safety factors for pile
capacity may be used as described in paragraph 4-2c. An iterative (nonlinear)
analysis of the pile group should be performed to determine that a state of
ductile, stable equilibrium is attainable even if individual piles will be
loaded to their peak, or beyond to their residual capacities. Special
provisions (such as field instrumentation, frequent or continuous field
monitoring of performance, engineering studies and analyses, constraints on
operational or rehabilitation activities, etc.) are required to ensure that
the structure will not catastrophically fail during or after extreme loading
conditions. Deviations from these criteria for extreme loading conditions
should be formulated in consultation with and approved by CECW-ED.

   (4) Foundation Properties. Determination of foundation properties is
partially dependent on types of loadings. Soil strength or stiffness, and
therefore pile capacity or stiffness, may depend on whether a load is vibratory, repetitive, or static and whether it is of long or short duration.

Soil-pile properties should, therefore, be determined for each type of loading to be considered.

c. Factor of Safety for Pile Capacity. The ultimate axial capacity, based on geotechnical considerations, should be divided by the following factors of safety to determine the design pile capacity for axial loading:

Method of Determining Capacity	Loading Condition	Minimum Factor of Safety Compression	Tension
Theoretical or empirical prediction to be verified by pile load test	Usual	2.0	2.0
	Unusual	1.5	1.5
	Extreme	1.15	1.15
Theoretical or empirical prediction to be verified by pile driving analyzer as described in Paragraph 5-4a	Usual	2.5	3.0
	Unusual	1.9	2.25
	Extreme	1.4	1.7
Theoretical or empirical prediction not verified by load test	Usual	3.0	3.0
	Unusual	2.25	2.25
	Extreme	1.7	1.7

The minimum safety factors in the table above are based on experience using the methods of site investigation, testing and analysis presented herein and are the basis for standard practice. Deviations from these minimum values may be justified by extensive foundation investigations and testing which reduce uncertainties related to the variability of the foundation material and soil strength parameters to a minimum. Such extensive studies should be conducted in consultation with and approved by CECW-ED. These minimum safety factors also include uncertainties related to factors which affect pile capacity during installation and the need to provide a design capacity which exhibits very little nonlinear load-deformation behavior at normal service load levels.

d. Allowable Stresses in Structural Members. Allowable design stresses for service loads should be limited to the values described in the following paragraphs. For unusual loadings as described in paragraph 4-2b(2), the allowable stresses may be increased by one third.

(1) Steel Piles. Allowable tension and compression stresses are given for both the lower and upper regions of the pile. Since the lower region of the pile is subject to damage during driving, the basic allowable stress should reflect a high factor of safety. The distribution of allowable axial tension or compression stress along the length of the pile is shown in Figure 4-1. This factor of safety may be decreased if more is known about the actual driving conditions. Pile shoes should be used when driving in dense sand strata, gravel strata, cobble-boulder zones, and when driving piles to refusal on a hard layer of bedrock. Bending effects are usually minimal in the lower region of the pile. The upper region of the pile may be subject to the effects of bending and buckling as well as axial load. Since damage in the upper region is usually apparent during driving, a higher allowable stress is permitted. The upper region of the pile is actually designed as a

beam-column, with due consideration to lateral support conditions. The allowable stresses for fully supported piles are as follows:

Tension or Compression in lower pile region

Concentric axial tension or compression only 10 kips per square inch ($1/3 \times F_y \times 5/6$)	10 kips per square inch (ksi) for A-36 material
Concentric axial tension or compression only with driving shoes ($1/3 \times F_y$)	12 ksi for A-36 material
Concentric axial tension or compression only with driving shoes, at least one axial load test and use of a pile driving analyzer to verify the pile capacity and integrity ($1/2.5 \times F_y$)	14.5 ksi for A-36 material

Combined bending and axial compression in upper pile region:

$$\left| \frac{f_a}{F_a} \pm \frac{f_{bx}}{F_b} \pm \frac{f_{by}}{F_b} \right| \leq 1.0$$

where

f_a = computed axial unit stress

F_a = allowable axial stress

$F_a = \dfrac{5}{6} \times \dfrac{3}{5} F_y = \dfrac{1}{2} F_y = 18$ ksi (for A-36 material)

f_{bx} and f_{by} = computed unit bending stress

F_b = allowable bending stress

$F_b = \dfrac{5}{6} \times \dfrac{3}{5} F_y = \dfrac{1}{2} F_y = 18$ ksi (for A-36 noncompact sections)

or

$F_b = \dfrac{5}{6} \times \dfrac{2}{3} F_y = \dfrac{5}{9} F_y = 20$ ksi (for A-36 compact sections)

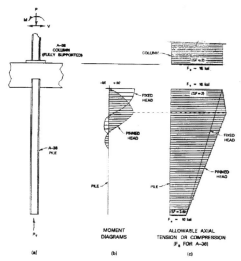

Figure 4-1. Allowable tension and compression
stress for steel piles

     For laterally unsupported piles the allowable stresses should be 5/6 of
the American Institute of Steel Construction (AISC) (Item 21) values for
beam-columns.

     (2) Concrete Piles. Design criteria for four types of concrete piles
(prestressed, reinforced, cast-in-place and mandrel driven) are presented in
the following paragraphs.

     (a) Prestressed Concrete Piles. Prestressed concrete piles are used
frequently and must be designed to satisfy both strength and serviceability
requirements. Strength design should follow the basic criteria set forth by
the American Concrete Institute (ACI) 318 (Item 19) except the strength reduction factor ($\emptyset$) shall be 0.7 for all failure modes and the load factor shall
be 1.9 for both dead and live loads. The specified load and strength reduction factors provide a safety factor equal to 2.7 for all combinations of dead
and live loads. To account for accidental eccentricities, the axial strength
of the pile shall be limited to 80 percent of pure axial strength, or the pile
shall be designed for a minimum eccentricity equal to 10 percent of the pile
width. Strength interaction diagrams for prestressed concrete piles may be
developed using the computer program CPGC (Item 16). Control of cracking in
prestressed piles is achieved by limiting the concrete compressive and tensile
stresses under service conditions to the values indicated in Table 4-1. The
allowable compressive stresses for hydraulic structures are limited to
approximately 85 percent of those recommended by ACI Committee 543 (Item 20)
for improved serviceability. Permissible stresses in the prestressing steel
tendons should be in accordance with Item 19. A typical interaction diagram,
depicting both strength and service load designs, is shown in Figure 4-2. The
use of concrete with a compressive strength exceeding 7,000 psi requires

Table 4-1

Allowable Concrete Stresses, Prestressed Concrete Piles

(Considering Prestress)

Uniform Axial Tension   0

Bending (extreme fiber)

   Compression $0.40 \ f_c'$

   Tension   0

For combined axial load and bending, the concrete stresses should be proportioned so that:

$$f_a + f_b + f_{pc} \leq 0.40 \ f_c'$$

$$f_a - f_b + f_{pc} \geq 0$$

Where:

    f_a = computed axial stress (tension is negative)

    f_b = computed bending stress (tension is negative)

    f_{pc} = effective prestress

    f_c' = concrete compressive strength

CECW-E approval. For common uses, a minimum effective prestress of 700 psi compression is required for handling and driving purposes. Excessively long or short piles may necessitate deviation from the minimum effective prestress requirement. The capacity of piles may be reduced by slenderness effects when a portion of the pile is free standing or when the soil is too weak to provide lateral support. Slenderness effects can be approximated using moment magnification procedures. The moment magnification methods of ACI 318, as modified by PCI, "Recommended Practice for the Design of Prestressed Concrete Columns and Walls" (Item 47), are recommended.

   (b) Reinforced Concrete Piles. Reinforced concrete piles shall be designed for strength in accordance with the general requirements of ACI 318 (Item 19) except as modified below. Load factors prescribed in ACI 318 should be directly applied to hydraulic structures with one alteration. The factored load combination "U" should be increased by a hydraulic load factor (H_f). This increase should lead to improved serviceability and will yield stiffer

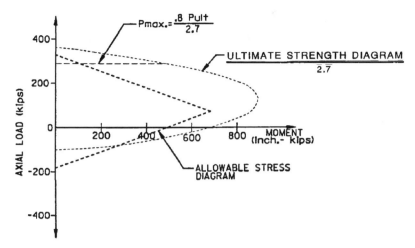

Figure 4-2. Typical interaction diagram, 16 × 16 in.
square prestressed concrete pile

members than those designed solely by ACI 318. The hydraulic load factor
shall be 1.3 for reinforcement calculations in flexure or compression, 1.65
for reinforcement in direct tension, and 1.3 for reinforcement in diagonal
tension (shear). The shear reinforcement calculation should deduct the shear
carried by the concrete prior to application of the hydraulic load factor. As
an alternate to the prescribed ACI load factors, a single load factor of 1.7
can be used. The 1.7 should then be multiplied by H_f. The axial compression
strength of the pile shall be limited to 80 percent of the ultimate axial
strength, or the pile shall be designed for a minimum eccentricity equal to
10 percent of the pile width. Strength interaction diagrams for reinforced
concrete piles may be developed using the Corps computer program CASTR
(Item 18). Slenderness effects can be approximated using the ACI moment
magnification procedures.

(c) Cast-in-Place and Mandrel-Driven Piles. For a cast-in-place pile,
the casing is top-driven without the aid of a mandrel, and the casing typically has a wall thickness ranging from 9 gage to 1/4 inch. The casing must
be of sufficient thickness to withstand stresses due to the driving operation
and maintain the cross section of the pile. The casing thickness for mandrel-
driven piles is normally 14 gage. Cast-in-place and mandrel-driven piles
should be designed for service conditions and stresses limited to those values
listed in Table 4-2. The allowable compressive stresses are reduced from
those recommended by ACI 543 (Item 20), as explained for prestressed concrete
piles. Cast-in-place and mandrel-driven piles shall be used only when full
embedment and full lateral support are assured and under conditions which
produce zero or small end moments, so that compression always controls. In
order for a pile to qualify as confined, the steel casing must be 14 gage
(US Standard) or thicker, be seamless or have spirally welded seams, have a
minimum yield strength of 30 ksi, be 17 inches or less in diameter, not be
exposed to a detrimental corrosive environment, and not be designed to carry a

Table 4-2

Cast-in-Place and Mandrel-Driven Piles, Allowable Concrete Stresses

(Participation of steel casing or shell disallowed)

Uniform Axial Compression	
Confined	$0.33\ f_c'$
Unconfined	$0.27\ f_c'$
Uniform Axial Tension	0
Bending (extreme fiber)	
Compression	$0.40\ f_c'$
Tension	0

For combined axial load and bending, the concrete stresses should be proportioned so that:

$$\left| \frac{f_a}{F_a} \pm \frac{f_b}{F_b} \right| \le 1.0$$

Where:

    f_a = computed axial stress

    F_a = allowable axial stress

    f_b = computed bending stress

    F_b = allowable bending stress

portion of the working load. Items not specifically addressed in this paragraph shall be in accordance with ACI 543.

(3) Timber Piles. Representative allowable stresses for pressure-treated round timber piles for normal load duration in hydraulic structures are:

Species	Compression Parallel to Grain (psi) F_a	Bending (psi) F_b	Horizontal Shear (psi)	Compression Perpendicular to Grain (psi)	Modulus of Elasticity (psi)
Pacific Coast (a)* Douglas Fir	875	1,700	95	190	1,500,000
Southern Pine (a)(b)*	825	1,650	90	205	1,500,000

(a) The working stresses for compression parallel to grain in Douglas Fir and Southern Pine may be increased by 0.2 percent for each foot of length from the tip of the pile to the critical section. For compression perpendicular to grain, an increase of 2.5 psi per foot of length is recommended.

(b) Values for Southern Pine are weighted for longleaf, slash, loblolly and shortleaf representatives of piles in use.

(c) The above working stresses have been adjusted to compensate for strength reductions due to conditioning and treatment. For untreated piles or piles that are air-dried or kiln-dried before pressure treatment, the above working stresses should be increased by dividing the tabulated values by the following factors:

Pacific Coast Douglas Fir:	0.90
Southern Pine:	0.85

(d) The allowable stresses for compression parallel to the grain and bending, derived in accordance with ASTM D2899, are reduced by a safety factor of 1.2 in order to comply with the general intent of Paragraph 13.1 of ASTM D2899 (Item 22).

(e) For hydraulic structures, the above values, except for the modulus of elasticity, have been reduced by dividing by a factor of 1.2. This additional reduction recognizes the difference in loading effects between the ASTM normal load duration and the longer load duration typical of hydraulic structures, and the uncertainties regarding strength reduction due to conditioning processes prior to treatment. For combined axial load and bending, stresses should be so proportioned that:

$$\left| \frac{f_a}{F_a} + \frac{f_b}{F_b} \right| \leq 1.0$$

where

   f_a = computed axial stress

   F_a = allowable axial stress

   f_b = computed bending stress

   F_b = allowable bending stress

   e. Deformations. Horizontal and vertical displacements resulting from applied loads should be limited to ensure proper operation and integrity of the structure. Experience has shown that a vertical deformation of 1/4 inch and a lateral deformation of 1/4 to 1/2 inch at the pile cap are representative of long-term movements of structures such as locks and dams. Operational requirements may dictate more rigid restrictions and deformations. For other structures such as piers, larger deformations may be allowed if the stresses in the structure and the piles are not excessive. Since the elastic spring constants used in the pile group analysis discussed later are based on a linear load versus deformation relationship at a specified deformation, it is important to keep the computed deformations at or below the specified value. Long-term lateral deformations may be larger than the computed values or the values obtained from load tests due to creep or plastic flow. Lateral deflection may also increase due to cyclic loading and close spacing. These conditions should be investigated when determining the maximum predicted displacement.

   f. Allowable Driving Stresses. Axial driving stresses calculated by wave equation analysis should be limited to the values shown in Figure 4-3.

   g. Geometric Constraints.

   (1) Pile Spacing. In determining the spacing of piles, consideration should be given to the characteristics of the soil and to the length, size, driving tolerance, batter, and shape of the piles. If piles are spaced too closely, the bearing value and lateral resistance of each pile will be reduced, and there is danger of heaving of the foundation, and uplifting or damaging other piles already driven. In general, it is recommended that end-bearing piles be spaced not less than three pile diameters on centers and that friction piles, depending on the characteristics of the piles and soil, be spaced a minimum of three to five pile diameters on center. Piles must be spaced to avoid tip interference due to specified driving tolerances. See paragraph 5-2a(3) for typical tolerances. Pile layouts should be checked for pile interference using CPGI, a program which is being currently developed and is discussed in paragraph 1-3c(b).

   (2) Pile Batter. Batter piles are used to support structures subjected to large lateral loads, or if the upper foundation stratum will not adequately resist lateral movement of vertical piles. Piles may be battered in opposite directions or used in combination with vertical piles. The axial load on a batter pile should not exceed the allowable design load for a vertical pile. It is very difficult to drive piles with a batter greater than 1 horizontal to 2 vertical. The driving efficiency of the hammer is decreased as the batter increases.

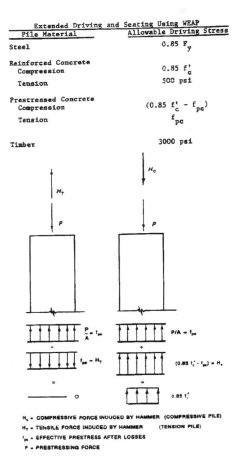

Extended Driving and Seating Using WEAP

Pile Material	Allowable Driving Stress
Steel	$0.85\ F_y$
Reinforced Concrete Compression	$0.85\ f_c'$
Tension	500 psi
Prestressed Concrete Compression	$(0.85\ f_c' - f_{pc})$
Tension	f_{pc}
Timber	3000 psi

H_c = COMPRESSIVE FORCE INDUCED BY HAMMER (COMPRESSIVE PILE)
H_T = TENSILE FORCE INDUCED BY HAMMER   (TENSION PILE)
f_{pc} = EFFECTIVE PRESTRESS AFTER LOSSES
P = PRESTRESSING FORCE

Figure 4-3.   Prestressed concrete pile
driving stresses

4-3.   <u>Pile Capacity</u>. Pile capacities should be computed by experienced
designers thoroughly familiar with the various types of piles, how piles behave when loaded, and the soil conditions that exist at the site.

a.   Axial Pile Capacity. The axial capacity of a pile may be represented
by the following formula:

$$Q_{ult} = Q_s + Q_t$$

$$Q_s = f_s A_s$$

$$Q_t = q A_t$$

where

Q_{ult} = ultimate pile capacity

Q_s = shaft resistance of the pile due to skin friction

Q_t = tip resistance of the pile due to end bearing

f_s = average unit skin resistance

A_s = surface area of the shaft in contact with the soil

q = unit tip-bearing capacity

A_t = effective (gross) area of the tip of the pile in contact with the soil

(1)   Piles in Cohesionless Soil.

(a)   Skin Friction.   For design purposes the skin friction of piles in sand increase linearly to an assumed critical depth (D_c) and then remain constant below that depth.   The critical depth varies between 10 to 20 pile diameters or widths (B), depending on the relative density of the sand.   The critical depth is assumed as:

D_c = 10B   for loose sands

D_c = 15B   for medium dense sands

D_c = 20B   for dense sands

The unit skin friction acting on the pile shaft may be determined by the following equations:

$$f_s = K\sigma'_v \tan \delta$$

$$\sigma'_v = \gamma'D \quad \text{for} \quad D < D_c$$

$$\sigma'_v = \gamma'D_c \quad \text{for} \quad D \geq D_c$$

$$Q_s = f_s A_s$$

where

K = lateral earth pressure coefficient (K_c for compression piles and K_t for tension piles)

σ'_v = effective overburden pressure

δ = angle of friction between the soil and the pile

4-11

γ' = effective unit weight of soil

D = depth along the pile at which the effective overburden pressure is calculated

Values of δ are given in Table 4-3.

Table 4-3

Values of δ

Pile Material	δ
Steel	0.67 ϕ to 0.83 ϕ
Concrete	0.90 ϕ to 1.0 ϕ
Timber	0.80 ϕ to 1.0 ϕ

Values of K for piles in compression (K_c) and piles in tension (K_t) are given in Table 4-4. Table 4-3 and Table 4-4 present ranges of values of δ and K based upon experience in various soil deposits. These values should be selected for design based upon experience and pile load test. It is not intended that the designer would use the minimum reduction of the ϕ angle while using the upper range K values.

Table 4-4

Values of K

Soil Type	K_c	K_t
Sand	1.00 to 2.00	0.50 to 0.70
Silt	1.00	0.50 to 0.70
Clay	1.00	0.70 to 1.00

Note: The above do not apply to piles that are prebored, jetted, or installed with a vibratory hammer. Picking K values at the upper end of the above ranges should be based on local experience. K , δ , and N_q values back calculated from load tests may be used.

For steel H-piles, A_s should be taken as the block perimeter of the pile and δ should be the average friction angles of steel against sand and sand against sand (ϕ). It should be noted that Table 4-4 is general guidance to be used unless the long-term engineering practice in the area indicates otherwise. Under prediction of soil strength parameters at load test sites has at times produced back-calculated values of K that exceed the values in Table 4-4. It has also been found both theoretically and at some test sites that the use of displacement piles produces higher values of K than does the use of nondisplacement piles. Values of K that have been used satisfactorily but with standard soil data in some locations are as follows in Table 4-5:

Table 4-5

Common Values for Corrected K

Soil Type	Displacement Piles		Nondisplacement Piles	
	Compression	Tension	Compression	Tension
Sand	2.00	0.67	1.50	0.50
Silt	1.25	0.50	1.00	0.35
Clay	1.25	0.90	1.00	0.70

Note: Although these values may be commonly used in some areas they should not be used without experience and testing to validate them.

(b)   End Bearing.   For design purposes the pile-tip bearing capacity can be assumed to increase linearly to a critical depth (D_c) and then remains constant.   The same critical depth relationship used for skin friction can be used for end bearing.   The unit tip bearing capacity can be determined as follows:

$$q = \sigma'_v N_q$$

where:

$$\sigma'_v = \gamma' D \quad \text{for} \quad D < D_c$$

$$\sigma'_v = \gamma' D_c \quad \text{for} \quad D \geq D_c$$

For steel H-piles A_t should be taken as the area included within the block perimeter.   A curve to obtain the Terzaghi-Peck (Item 59) bearing capacity factor N_q (among values from other theories) is shown in Figure 4-4.   To use the curve one must obtain measured values of the angle of internal friction (ϕ) which represents the soil mass.

(c)   Tension Capacity.   The tension capacity of piles in sand can be calculated as follows using the K values for tension from Table 4-4:

$$Q_{ult} = Q_{s_{tension}}$$

(2)   Piles in Cohesive Soil.

(a)   Skin Friction.   Although called skin friction, the resistance is due to the cohesion or adhesion of the clay to the pile shaft.

$$f_s = c_a$$

$$c_a = \alpha c$$

$$Q_s = f_s A_s$$

4-13

where

    c_a = adhesion between the clay and the pile

    α = adhesion factor

    c = undrained shear strength of the clay from a Q test

The values of α as a function of the undrained shear are given in Figure 4-5a.

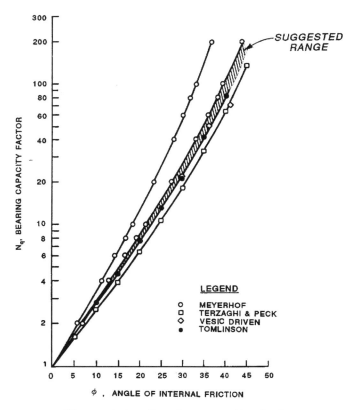

Figure 4-4. Bearing capacity factor

An alternate procedure developed by Semple and Rigden (Item 56) to obtain values of α which is especially applicable for very long piles is given in Figure 4-5b where:

$$\alpha = \alpha_1 \alpha_2$$

and

$$f_s = \alpha c$$

4-14

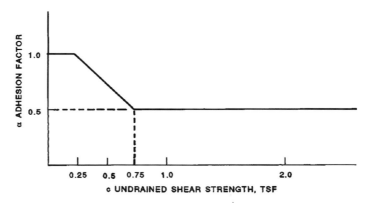

Figure 4-5a. Values of α versus undrained shear strength

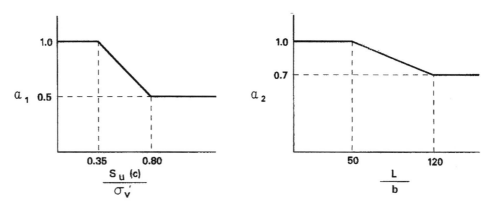

Figure 4-5b. Values of $\alpha_1 \alpha_2$ applicable for very long piles

(b) End Bearing. The pile unit-tip bearing capacity for piles in clay can be determined from the following equation:

$$q = 9c$$

$$Q_t = A_t q$$

However, the movement necessary to develop the tip resistance of piles in clay soils may be several times larger than that required to develop the skin friction resistance.

(c) Compression Capacity. By combining the skin friction capacity and the tip bearing capacity, the ultimate compression capacity may be found as follows:

$$Q_{ult} = Q_s + Q_t$$

(d) Tension Capacity. The tension capacity of piles in clay may be calculated as:

$$Q_{ult} = Q_s$$

(e) The pile capacity in normally consolidated clays (cohesive soils)
should also be computed in the long-term S shear strength case. That is,
develop a S case shear strength trend as discussed previously and proceed as
if the soil is drained. The computational method is identical to that
presented for piles in granular soils, and to present the computational
methodology would be redundant. It should be noted however that the shear
strengths in clays in the S case are assumed to be $\phi > 0$ and $C = 0$.
Some commonly used S case shear strengths in alluvial soils are as follows
in Table 4-6:

Table 4-6

S Case Shear Strength

Soil Type	Consistency	Angle of Internal Friction ϕ
Fat clay (CH)	Very soft	13° to 17°
Fat clay (CH)	Soft	17° to 20°
Fat clay (CH)	Medium	20° to 21°
Fat clay (CH)	Stiff	21° to 23°
Silt (ML)		25° to 28°

Note: The designer should perform testing and select shear
       strengths. These general data ranges are from test on
       specific soils in site specific environments and may not
       represent the soil in question.

(3) Piles in Silt.

(a) Skin Friction. The skin friction on a pile in silt is a two component resistance to pile movement contributed by the angle of internal friction
(ϕ) and the cohesion (c) acting along the pile shaft. That portion of the resistance contributed by the angle of internal friction (ϕ) is as with the sand
limited to a critical depth of (D_c), below which the frictional portion
remains constant, the limit depths are stated below. That portion of the
resistance contributed by the cohesion may require limit if it is sufficiently
large, see Figures 4-5a and b. The shaft resistance may be computed as
follows:

$$K\gamma'D \tan \delta + \alpha c$$

$$\text{where } (D \leq D_c)$$

$$Q_s = A_s f_s$$

4-16

where

Q_s = capacity due to skin resistance

f_s = average unit skin resistance

A_s = surface area of the pile shaft in contact with soil

K = see Table 4-4

α = see Figures 4-5a and b

D = depth below ground up to limit depth D_c

δ = limit value for shaft friction angle from Table 4-3

(b) End Bearing. The pile tip bearing capacity increases linearly to a critical depth (D_c) and remains constant below that depth. The critical depths are given as follows:

$$D_c = 10 \text{ B for loose silts}$$

$$D_c = 15 \text{ B for medium silts}$$

$$D_c = 20 \text{ B for dense silts}$$

The unit and bearing capacity may be computed as follows:

$$q = \sigma'_v N_q$$

$$\sigma'_v = \gamma' D \quad \text{for} \quad D < D_c$$

$$\sigma'_v = \gamma' D_c \quad \text{for} \quad D \geq D_c$$

$$Q_t = A_t q$$

where

N_q = Terzaghi bearing capacity factor, Figure 4-4

σ'_v = vertical earth pressure at the tip with limits

A_t = area of the pile tip, as determined for sands

(c) Compression Capacity. By combining the two incremental contribu-
tors, skin friction and end bearing the ultimate capacity of the soil/pile may
be computed as follows:

$$Q_{ult} = Q_s + Q_t$$

(d) Tension Capacity. The tension capacity is computed by applying the
appropriate value of K_t from Table 4-4 to the unit skin friction equation
above.

$$Q_{ult} = Q_{s_{tension}}$$

(e) It is recommended that when designing pile foundations in silty
soils, considerations be given to selecting a very conservative shear strength
from classical R shear tests. It is further recommended that test piles be
considered as a virtual necessity, and the possibility that pile length may
have to be increased in the field should be considered.

(4) Piles in Layered Soils. Piles are most frequently driven into a
layered soil stratigraphy. For this condition, the preceding methods of
computation may be used on a layer by layer basis. The end bearing capacity
of the pile should be determined from the properties of the layer of soil
where the tip is founded. However, when weak or dissimilar layers of soil
exist within approximately 5 feet or 8 pile tip diameters, whichever is the
larger, of the tip founding elevation the end bearing capacity will be
affected. It is necessary to compute this affect and account for it when
assigning end bearing capacity. In computing the skin resistance, the
contribution of each layer is computed separately, considering the layers
above as a surcharge and applying the appropriate reduction factors for the
soil type within that increment of pile shaft.

(a) Skin Friction. The skin friction contributed by different soil
types may be computed incrementally and summed to find the ultimate capacity.
Consideration should be given to compatibility of strain between layers when
computing the unit skin resistance.

$$Q_s = \sum_{i=1}^{N} f_{s_i} A_{s_i}$$

where

f_{s_i} = unit skin resistance in layer i

A_{s_i} = surface area of pile in contact with layer i

N = total number of layers

(b) End Bearing. The pile tip bearing should be computed based upon the soil type within which the tip is founded, with limits near layer boundaries mentioned above. Using the overlying soil layers as surcharge the following equations may be used.

$$\text{Sand or Silt:} \qquad q = \sigma'_v N_q$$

$$\sigma'_v = \gamma' D \quad \text{for} \quad D < D_c$$

$$\sigma'_v = \gamma' D_c \quad \text{for} \quad D > D_c$$

$$Q_t = A_t q$$

$$\text{Clay:} \qquad q = 9c$$

$$Q_t = A_t q$$

(c) Compression Capacity. By combining the skin resistance and end bearing, the ultimate capacity of the soil/pile may be computed as follows:

$$Q_{ult} = Q_s + Q_t$$

(d) Tension Capacity. The tension capacity may be computed by applying the appropriate values of K_t from Table 4-4 as appropriate for granular soils to the incremental computation for each layer and then combining to yield:

$$Q_{ult} = Q_{s_{tension}}$$

(5) Point Bearing Piles. In some cases the pile will be driven to refusal upon firm good quality rock. In such cases the capacity of the pile is governed by the structural capacity of the pile or the rock capacity.

(6) Negative Skin Friction.

(a) Negative skin friction is a downward shear drag acting on piles due to downward movement of surrounding soil strata relative to the piles. For such movement of the soils to occur, a segment of the pile must penetrate a compressible soil stratum that consolidates. The downward drag may be caused by the placement of fill on compressible soils, lowering of the groundwater table, or underconsolidated natural or compacted soils. The effect of these occurrences is to cause the compressible soils surrounding the piles to consolidate. If the pile tip is in a relatively stiff soil, the upper compressible stratum will move down relative to the pile, inducing a drag load. This load can be quite large and must be added to the structural load for purposes of assessing stresses in the pile. Vesic (Item 60) stated that a relative downward movement of as little as 0.6 inch of the soil with respect to the pile may be sufficient to mobilize full negative skin friction. The geotechnical capacity of the pile is unaffected by downdrag, however downdrag does serve to increase settlement and increase the stresses in the pile and pile cap.

(b) For a pile group, it can be assumed that there is no relative movement between the piles and the soil between the piles. Therefore, the total force acting down is equal to the weight of the block of soil held between the piles, plus the shear along the pile group perimeter due to negative skin friction. The average downward load transferred to a pile in a pile group Q_{nf} can be estimated by

$$Q_{nf} = \frac{1}{N} [A\gamma L + sLP] \tag{1}$$

where

A = horizontal area bounded by the pile group (cross-sectional area of piles and enclosed soil)

N = number of piles in pile group

γ = unit weight of fill or compressible soil layers

L = length of embedment above the bottom of the compressible soil layers

s = shear resistance of the soil

P = perimeter of the area A

(c) For a single pile, the downward load transferred to the pile is equal to the shearing resistance along the pile as shown in Equation 2.

$$Q_{nf} = sLP' \tag{2}$$

where P' = perimeter of pile. The total applied load (Q_T) on a pile group or single pile is the live load, dead load, and the drag load due to negative skin friction.

$$Q_T = Q + A\gamma L + sLP \quad \text{(pile group)} \tag{3a}$$

$$Q_T = Q + sLP' \quad \text{(single pile)} \tag{3b}$$

where Q = live load plus dead load.

(d) Commentary. Equation 1 for pile groups was used by Teng (Item 58) and Terzaghi and Peck (Item 59). However, in Peck, Hanson, and Thornburn (Item 46), the shear resistance on the perimeter was eliminated. Both Teng and Terzaghi and Peck state that the component due to shear resistance is the larger value. Teng recommends using the lesser of the summation of shear resistance for individual piles of a pile group and Equation 1. Bowles (Item 27) and the Department of the Navy, Naval Facilities Engineering Command (NAVFAC) (Item 33) both use a coefficient relating the overburden pressures to the shearing resistance around the pile. NAVFAC gives different values for clay, silt, and sands and references Garlanger (Item 35), Prediction of Downdrag Load at the Cutler Circle Bridge. Bowles uses the block perimeter resistance for a pile group similar to Equation 1. Bowles recommends using the higher value of Equation 1 and, between the summation of shear resistance on a single pile, using the coefficient relating overburden pressure to shear resistance and Equation 1. NAVFAC does not use the block perimeter resistance for a pile group. For single piles, NAVFAC uses the coefficient times the effective vertical stress.

b. Pile Group Capacity. The pile group capacity for piles in cohesionless and cohesive soils is given below.

(1) Piles in Cohesionless Soil. The pile group efficiency η is defined as:

$$\eta = \frac{Q_{group}}{NQ_{ult}}$$

where

Q_{group} = ultimate capacity of a pile group

N = number of piles in a group

Q_{ult} = ultimate capacity of a single pile

The ultimate group capacity of driven piles in sand is equal to or greater than the sum of the ultimate capacity of the single piles. Therefore in practice, the ultimate group capacity of driven piles in sand not underlain by a weak layer, should be taken as the sum of the single pile capacities ($\eta = 1$). For piles jetted into sand, η is less than one. For piles underlain by a weak layer, the ultimate group capacity is the smaller of (a) the sum of the single pile ultimate capacities or (b) the capacity of an equivalent pier with the geometry defined by enclosing the pile group (Item 59). The base strength should be that of the weak layer.

(2) Piles in Cohesive Soil. The ultimate group capacity of piles in clay is the smaller of (a) the sum of the single pile ultimate capacities or (b) the capacity of an equivalent pier (Item 59). The ultimate group capacity of piles in clay is given by the smaller of the following two equations:

$$Q_{group} = NQ_{ult}$$

$$Q_{group} = 2(B_g + L_g)D\bar{c} + \left[5\left(1 + \frac{D}{5B_g}\right)\left(1 + \frac{B_g}{5L_g}\right) \right] c_b L_G B_G$$

where

$$N_c = 5\left(1 + \frac{D}{5B_g}\right)\left(1 + \frac{B_g}{5L_g}\right) \leq 9$$

and:

B_g = width of the pile group

L_g = length of the pile group

D = depth of the pile group

$\overline{c}$ = weighted average of undrained shear strength over the depth of pile embedment. $\overline{c}$ should be reduced by α from Figure 4-5.

c_b = undrained shear strength at the base of the pile group

This equation applies to a rectangular section only. It should be modified for other shapes.

4-4. Settlement. The load transfer settlement relationship for single piles and pile groups is very complex. Most settlement analysis methods are based on empirical methods and give only a rough approximation of the actual settlement. However, settlements of single piles and pile groups should be calculated to give the designer a perception of how the structure will perform and to check that these calculated settlements are within acceptable limits (paragraph 4-2e). Calculated foundation settlements should be compatible with the force-movement relationships used in designing the structure.

a. Single Piles.

(1) Semi-Empirical Method. The semi-empirical method for calculating the settlement of single piles is the method proposed by Vesic (Item 60). The settlement of a single pile is given by the equation

$$w = w_s + w_{pp} + w_{ps}$$

where

w = vertical settlement of the top of a single pile

w_s = amount of settlement due to the axial deformation of the pile shaft

w_{pp} = amount of settlement of the pile tip due to the load transferred at the tip

w_{ps} = amount of settlement of the pile tip caused by load transmitted along the pile shaft

The axial deformation of the pile shaft is given by:

$$w_s = (Q_p + \alpha_s Q_s)\frac{L}{AE}$$

4-22

where

Q_p = tip resistance of the pile for the design load for which the
settlement is being calculated

α_s = number that depends on the skin friction distribution along the pile
(Figure 4-6)

Q_s = shaft resistance of the pile for the design load for which the
settlement is being calculated

L = length of the pile

A = cross-sectional area of the pile

E = modulus of elasticity of the pile

Lesser values of α_s have been observed in long driven piles subject to hard
driving. A typical value for piles driven into dense sand may be around 0.1.
Lesser values of α_s are also observed for long, flexible friction piles
where under working loads, only a fraction of the shaft length transmits load.
The settlement at the tip of the pile can be calculated by the following
equations:

$$w_{pp} = \frac{C_p Q_p}{Bq}$$

$$w_{ps} = \frac{C_s Q_s}{Dq}$$

where

C_p = empirical coefficient given in Table 4-7

B = pile diameter or width

q = unit ultimate tip bearing capacity

C_s = coefficient given by the following equation
$$C_s = (0.93 + 0.16 \sqrt{D/B}) \ C_p$$

D = embedded pile length

The values of C_p given in Table 4-7 are for long-term settlement of the pile
where the bearing stratum beneath the pile tip extends a minimum of 10B
beneath the pile tip and where such soil is of equal or higher stiffness than
that of the soil at the tip elevation. The value of C_p will be lower if
rock exists nearer the pile tip than 10B. If rock exists at 5B beneath the
pile tip, use 88 percent of w_{pp} in the settlement calculations. If rock

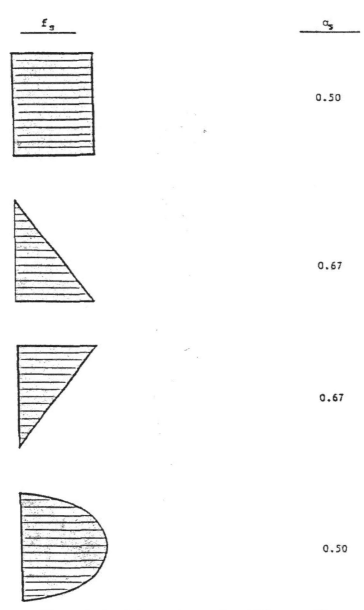

Figure 4-6.   Values of α_s for different skin
friction distributions (Item 60)

exists at 1B beneath the pile tip, use 51 percent of w_{pp} in the settlement calculations. Unless a highly compressible layer exists beneath the pile tip, consolidation settlement should not be significant and normally does not exceed 15 percent of the total settlement. If a highly compressible layer does exist beneath the pile tip, a consolidation-settlement analysis should be performed to determine the additional long-term settlement that will occur.

Table 4-7

Value of C_p

Soil Type	Driven Piles	Bored Piles
Sand (dense to loose)	0.02 to 0.04	0.09 to 0.18
Clay (stiff to soft)	0.02 to 0.03	0.03 to 0.06
Silt (dense to loose)	0.03 to 0.05	0.09 to 0.12

(2) Elastic Method. For the elastic method of calculating single pile settlement, the designer is referred to Item 48.

(3) t-z Curve Methods. The t-z curve methods of calculating settlement of a single pile requires the use of a computer program and t-z curves (load transfer relationships for the pile-soil system). A number of computer programs are available from WES (Items 4, 13) for performing t-z curve analyses. Various load-transfer relationships (t-z curves) exist in supplemental literature (Items 9, 28, 29, 38, and 61).

b. Pile Groups. A number of methods exist for calculating settlement of groups of piles. The designer of a pile foundation should be aware that less is known about settlement of pile groups than any item discussed in this section.

(1) Group Settlement Factors. The simplest method for calculating settlement of a group of piles implements a group settlement factor.

$$S = \zeta_g w$$

where

S = settlement of a group of piles

ζ_g = group settlement factor

w = settlement of a single pile

The simplest expression for the group settlement factor is:

$$\zeta_g = \left(\frac{\bar{B}}{B}\right)^{0.5}$$

where

   $\overline{B}$ = width of the pile group

   B = diameter or width of a single pile

Three things must be kept in mind when using the above method:

   (a)   It is an approximate method.

   (b)   The group settlement factor was determined empirically from pile groups in sand.

   (c)   The settlement of a pile group is larger than that of a single pile with the same load per pile.   This method takes that fact into account.

The following expression for the group settlement factor has been used for pile groups in clay:

$$\zeta_g = 1 + \sum_{i=1}^{n} \frac{B_i}{\pi S_i}$$

where

   N = number of piles in group

   s_i = distance from pile i to the location in the group where the group settlement is to be calculated

   (2)   Empirical Method.   The empirical method for calculating the settlement of a group of piles is the method presented in Item 40.   It is based on the concept that the pile group can be treated as an equivalent pier.   For a group of friction piles, the equivalent footing is assumed to be founded at an effective depth of two-thirds of the pile embedment in the bearing stratum. For a group of end bearing piles, the equivalent footing is assumed to be founded at the pile tips.

   (a)   Groups in Sand.   For calculating the settlement of pile groups in a homogeneous sand deposit not underlain by a more compressible soil at greater depth, the following expressions can be used:

$$S = \frac{2p\ \overline{B}}{\overline{N}}\ I$$

$$I = 1 - \frac{D'}{8\overline{B}} \geq 0.5$$

where

S = settlement of the pile group in inches

p = net foundation pressure, is defined as the applied load divided by the horizontal area of the group in tons per square foot

$\overline{B}$ = width of the pile group in feet

I = influence factor of effective group embedment

$\overline{N}$ = average corrected standard penetration resistance in blows per foot within the zone of settlement (extending to a depth equal to the pile group width beneath the pile tip in homogeneous soil)

D' = embedment depth of the equivalent pier

In using the above equation, the measured blow counts should be corrected to an effective overburden pressure of 1 ton per square foot as suggested in Item 46. The calculated value of settlement should be doubled for silty sand.

(b) Groups in Clay. The settlement of pile groups in clay or above a clay layer can be estimated from the initial deformation and consolidation properties of the clay. The pile group is treated as an equivalent pier and allowance is made for the effective foundation embedment and compressible stratum thickness as outlined above. (PHT reference)

(3) Elastic Methods. For the elastic methods of calculating settlements of a pile group, the designer is referred to Item 48. These methods require the estimation of a secant modulus.

(4) Stiffness Method. The stiffness method of analysis of pile groups as outlined in paragraph 4-5 can be used to calculate settlements of pile groups. As mentioned for other methods for calculating settlement of pile groups in clay, this method does not take into account any consolidation of the clay and must be corrected for settlement due to consolidation if such consolidation occurs.

4-5. Pile Group Analysis.

a. General. Several approximate methods for analysis of pile groups have been used. These graphical or numerical methods distribute applied loads to each pile within the group based on pile location, batter, and cross-sectional area. These approaches did not consider lateral soil resistance, pile stiffness, pile-head fixity, structure flexibility, or any effects of pile-soil interaction. Such factors significantly affect the distribution of forces among the piles and, if ignored, can result in an unconservative and erroneous pile design. Therefore, these methods should not be used except for very simple, two-dimensional (2-D) structures where the lateral loads are small (less than 20 percent of the vertical loads).

b. Stiffness Methods.

(1) General. The behavior of the structure-pile-soil system is non-linear. However, it is not practical to apply nonlinear theory to the

analysis and design of large pile groups in a production mode. Therefore, it is necessary to develop elastic constants which satisfactorily represent the nonlinear, nonelastic behavior. An approach for pile group analysis using force-displacement relationships has been developed. This method, referred to as the stiffness method, accounts for all of the variables mentioned above. The stiffness method is based on work published by A. Hrennikoff (Item 37). This method involves representation of individual pile-soil behavior by axial, lateral, rotational, and torsional stiffness constants. Individual pile forces are equal to the corresponding pile displacements times the pile-soil stiffness. Hrennikoff's analysis was restricted to two dimensions and piles with identical properties. Aschenbrenner (Item 26) extended the solution to three dimensions, and Saul (Item 54) used matrix methods to incorporate position and batter of piles, and piles of different sizes, and materials. The Saul approach is the basis for the pile analysis presented in the following paragraphs. These stiffness methods should be used for the analysis and design of all but the simplest pile groups. This method is implemented in the computer program CPGA (Item 5).

(2) Pile-Soil Model. In the stiffness method of pile analysis, the structure is supported by sets of six single degree-of-freedom springs which are attached to the base of the structure. These springs represent the action of the pile-soil foundation when the structure is displaced due to applied forces (Figure 4-7). The pile-load springs are linearly elastic and are used

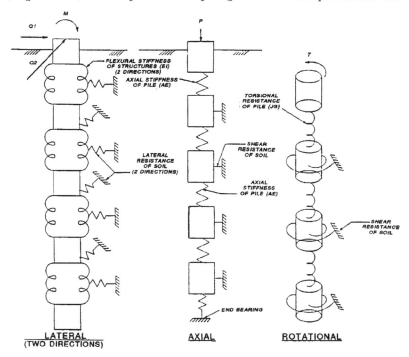

Figure 4-7.   Spring model of pile-soil interaction

4-28

to account for all the variables and nonlinearities of the foundation. The behavior of each pile is represented by spring (or stiffness) constants in matrix form:

$$\{q\}_i = [B]_i \{u\}_i$$

where

$$(q)_i = \begin{Bmatrix} F_1 \\ F_2 \\ F_3 \\ M_1 \\ M_2 \\ M_3 \end{Bmatrix} = \text{pile forces and moments in i}^{\text{th}} \text{ pile}$$

$[B]_i$ = matrix of stiffness constants for i^{th} pile

$$(\mu)_i = \begin{Bmatrix} U \\ V \\ W \\ \Theta_1 \\ \theta_2 \\ \Theta_3 \end{Bmatrix} = \text{displacements and rotations of i}^{\text{th}} \text{ pile}$$

The total foundation stiffness is the summation of all the individual pile stiffnesses assembled into a global foundation stiffness matrix:

$$[K] = \sum_{i=1}^{n} [K]_i$$

where

[K] = total pile group stiffness

n = number of piles in the foundation

$[K]_i$ = stiffness of ith pile transformed to global coordinates

The elastic response of each pile to applied forces is based on a subgrade reaction assumption. This assumption is that the lateral resistance of the soil to pile displacements can be modeled as a series of linear springs connected to an individual pile. Therefore, the behavior of each set of springs is affected only by the properties of the pile and the surrounding soil and not by the behavior of adjacent piles. This approximation is necessary for computational simplicity and to allow for easy adaptability of the model to complications such as changes in soil type. Analytical results have been compared to actual field results from pile load tests for numerous cases and have demonstrated that this pile-soil model is satisfactory for the analysis of pile groups at working load. However, the designer should always be aware of the model limitations. A more realistic approach is being developed for design. Methods for determining the stiffness constants are presented in paragraph 4-5c.

(3) Rigid Base Versus Flexible Base. Distribution of the loads applied by the structure to each pile is affected by many factors. One important assumption is related to the flexibility of the pile cap. The pile cap (structure) can be modeled as a rigid or a flexible body. If the structure is assumed to behave as a rigid body, then the stiffness of the pile cap is infinite relative to the stiffness of the pile-soil system. For a rigid pile cap deformations within the structure are negligible, and the applied loads are distributed to each pile on the basis of rigid body behavior (Figure 4-8) as is the case in CPGA (Item 5). If the pile cap is assumed to be a flexible body, then the internal deformations of the structure are also modeled and play an important role in the distribution of the applied loads to each pile (Figure 4-9). When performing a pile group analysis, one of the first design decisions that must be made is how to model the flexibility of the structure. Parametric studies should be performed to determine the effects of the structure stiffness on the pile forces. For example, a pile-founded dam pier could be idealized using a 2-D beam element for the structure and springs for the piles. Available computer programs (such as SAP, STRUDL, CFRAME, etc.) can be used to vary the stiffness of the beams (structure), and the axial and lateral stiffness of the springs (piles), and thereby determine which pile cap assumption is appropriate. For either type of the pile cap, the piles are modeled as linear elastic springs.

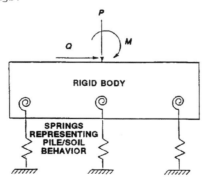

Figure 4-8. Rigid pile cap on a spring (pile) foundation

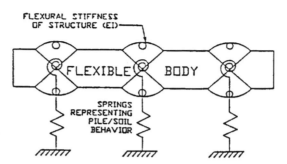

FLEXURAL STIFFNESS
OF STRUCTURE (EI)

FLEXIBLE BODY

SPRINGS
REPRESENTING
PILE/SOIL
BEHAVIOR

Figure 4-9. Flexible pile cap on
spring (pile)
foundation

(4) Nonlinear Effects. A pile group analysis is normally a linear elastic model. Actual load-deflection relationships for the pile-soil system can
be nonlinear. Programs such as PILGP2R have been developed for nonlinear
analysis of pile groups. The major disadvantage with using nonlinear pile
group analysis programs is that they can only be used to analyze small pile
groups, 30 piles or less. Many pile groups for hydraulic structures consist
of 200 piles or more. Linear elastic pile group programs can approximate
satisfactorily the nonlinear group analysis programs at working loads. A
comparison was conducted for two typical pile groups using PILGP2R to perform
a nonlinear analysis and using CPGA to perform a linear analysis which
approximates nonlinear behavior (Item 45). The results for these two pile
groups were in good agreement. The following methods for choosing stiffness
coefficients are used to perform the linear CPGA analyses which approximately
model nonlinear behavior.

c. Soil and Pile Properties. The soil-pile stiffness is a function of
the pile structural properties, soil properties, degree of pile restraint
against rotation, and pile-head movement. The pile properties needed to determine the spring stiffnesses are the modulus of elasticity, moment of
inertia, cross-sectional area, width, and length. The soil properties needed
to determine the spring stiffnesses are the undrained shear strength or angle
of internal friction, and the unit weight. An estimate of pile-head movement
is needed to determine the linear spring stiffnesses. This is accomplished by
using a secant modulus corresponding to an estimated pile-head movement. If
the calculated pile head movements reasonably agree with the estimated values,
then the solution is acceptable; if not, then a new estimate of pile head
movements must be used. (See paragraph 4-2e for additional discussion.)

d. Axial Stiffness. The axial pile stiffness is expressed as:

$$b_{33} = C_{33} \frac{AE}{L}$$

where

b$_{33}$ = axial pile stiffness

C$_{33}$ = constant which accounts for the interaction between the soil and the pile

A = cross-sectional area of the pile

E = modulus of elasticity of the pile

L = length of the pile

The term AE/L is the elastic stiffness of the pile acting as a short column with no soil present. The coefficient (C$_{33}$) accounts for the stiffness of the soil-pile system. The relationship between axial load capacity, movements of the pile head and tip, and load transfer along the shaft of friction piles is presented in the companion volume "Theoretical Manual for the Design of Pile Foundations," which is currently in preparation and is discussed in paragraph 1-3c(10).

(1) For design purposes, C$_{33}$ for a compression pile ranges between 1.0 and 2.0 although values as low as 0.1 and as high as 3.0 have been noted in the literature. There appears to be a relationship between C$_{33}$ and pile length. Longer piles tend to have higher values of C$_{33}$ than shorter piles. C$_{33}$ for tension piles in sand can be taken as one half of the value used for compression piles. For tension piles in clay use 75 to 80 percent of the value of C$_{33}$ for compression piles.

(2) Long-term loading, cyclic loading, pile group effects, and pile batter can affect C$_{33}$. In sand, long-term loading has little effect on the value of C$_{33}$; however consolidation in clay due to long-term loading can reduce C$_{33}$. At present, the effect of cyclic loading on C$_{33}$ is neglected. For design purposes, if piles are driven to refusal in sand or to a hard layer, there is no change in the value of C$_{33}$ for pile groups; however, C$_{33}$ may be reduced for groups of friction piles.

(3) The value of C$_{33}$ for single piles can be calculated using the following equation:

$$C_{33} = \frac{\Delta}{\delta}$$

where

$$\Delta = \frac{PL}{AE}$$

δ = axial movement of the pile head due to axial load P

P = allowable axial design load for the pile

For axial stiffness, the load-deflection curve is essentially linear to one-half of the ultimate pile capacity (the design load), so nonlinearity of the axial pile stiffness can be neglected. Methods for calculating C_{33} from the above equations include empirical methods (Item 60), Winkler foundation analysis (Item 55), t-z curve analyses (Items 9, 28, 29, 38, and 61), finite element methods, and elastic method (Item 48). Values of C_{33} can be determined most accurately from pile load tests, where C_{33} can be determined to approximate the linear portion of the pile load-deflection curve.

e. Lateral Stiffness. Expressions for lateral pile stiffness are given in Item 17. The lateral pile stiffness expressions contain the following terms:

E = modulus of elasticity of the pile material

I = moment of inertia of the pile section

C_1 = pile head-cap fixity constant (rotational restraint between pile head and pile cap)

E_s = modulus of horizontal subgrade reaction (expressed as soil reaction per unit length of pile per unit of lateral deflection)

n_h = constant of horizontal subgrade reaction (linear variation of E_s with depth i.e., $E_s = n_h x$)

Lateral pile stiffness expressions containing E_s (modulus of horizontal subgrade reaction not a function of depth) are assumed constant for overconsolidated clays. Lateral pile stiffness expressions containing n_h (modulus of horizontal subgrade reaction increasing linearly with depth) are used for sands and normally consolidated clays. Since the upper portion (10 pile diameters or less) of the soil profile usually controls the behavior of laterally loaded piles, most onshore clay deposits can be represented with a constant modulus of horizontal subgrade reaction. E_s and n_h are not constants. They both vary with deflection of the pile head. This is due to the fact that linear lateral stiffnesses are used to represent a nonlinear problem. To determine appropriate values of E_s or n_h, an estimate of lateral deflection must be made. If the calculated values of lateral deflection match the estimated values, then the correct value of E_s or n_h was used in the analysis. If not, a new value of E_s or n_h must be used based on the calculated deflection. For design, ranges of E_s or n_h are used to take into account variation of pile properties in different directions, variation of lateral pile deflection caused by different loading conditions, and variation of soil properties. After the analyses are completed, the calculated lateral deflection should be checked to make sure they correspond to the range of values of E_s or n_h assumed. If they do not, then the assumed range should be modified.

(1) Calculation of E_s or n_h. The first step in determining the range of E_s or n_h values to use in pile design is to determine curves of the variation of E_s or n_h with lateral pile head deflection. These curves can be estimated from plots of pile-head deflection versus applied lateral load (load-deflection curves). The pile-head, load-deflection curves can be obtained from lateral pile load tests or from p-y curve analyses (Items 13,

39, 50, 51, 52, and 53). From the load-deflection curves, the variation of E_s or n_h with deflection can be obtained using these equations for the case of applied groundline shear and zero applied moment.

$$n_h = \frac{C_n \left(\dfrac{P_t}{Y_t}\right)^{1.67}}{(EI)^{0.67}}$$

or

$$E_s = \frac{C_E \left(\dfrac{P_t}{Y_t}\right)^{1.33}}{(EI)^{0.33}}$$

where

C_n = 0.89 for a fixed-head pile or 4.41 for a free-head pile

P_t = lateral load applied at the top of the pile at the ground surface

Y_t = lateral deflection of the top of the pile at the ground surface

C_E = 0.63 for a fixed-head pile or 1.59 for a free-head pile

Use consistent units C_n and C_E are nondimensional constants.

(2) Stiffness Reduction Factors. Values of n_h or E_s calculated as outlined in the preceding paragraphs are for a single pile subject to static loading. Groups of piles, cyclic loading, and earthquake loading cause a re-duction in E_s and n_h. Reducing E_s and n_h increases the pile deflec-tions and moments at the same load level. The value of E_s or n_h for a single pile is divided by a reduction factor (R) to get the value of E_s or n_h for groups of piles, cyclic loading, or earthquake loading.

(a) Group Effects. Laterally loaded groups of piles deflect more than a single pile loaded with the same lateral load per pile as the group. This increased deflection is due to overlapping zones of stress of the individual piles in the group. The overlapping of stressed zones results in an apparent reduction in soil stiffness. For design, these group effects are taken into account by reducing the values of E_s or n_h by a group reduction factor (R_g). The group reduction factor is a function of the pile width (B), pile spacing, and number of piles in the group. Pile groups with center-line-to-center-line pile spacing of 2.5B perpendicular to the direction of loading and 8.0B in the direction of loading have no reduction in E_s or n_h. The group reduction factors for pile groups spaced closer than mentioned above are:

Center-Line-to-Center-Line Pile Spacing in Direction of Loading	Group Reduction Factor R_g
3B	3.0
4B	2.6
5B	2.2
6B	1.8
7B	1.4
8B	1.0

More recent data from pile group tests (Item 1, 8, and 12) suggest that these values are conservative for service loads, but at the present time no new procedure has been formalized.

(b) Cyclic Loading Effects. Cyclic loading of pile foundations may be due to tide, waves, or fluctuations in pool. Cyclic loading causes the deflection and moments of a single pile or a group of piles to increase rapidly with the number of cycles of load applied up to approximately 100 cycles, after which the deflection and moments increase much more slowly with increasing numbers of cycles. In design, cyclic loading is taken into account by reducing the values of E_s or n_h by the cyclic loading reduction factor (R_c). A cyclic loading reduction factor of 3.0 is appropriate for preliminary design.

(c) Combined Effect, Group and Cyclic Loading. When designing for cyclic loading of a group of piles, E_s or n_h for a single, statically loaded pile is divided by the product of R_g and R_c.

(d) Earthquake Loading Effects. The loading on the foundation induced by a potential earthquake must be considered in seismic active areas. The designer should first consider probability of an earthquake occurring during the life of the structure. If there is a likelihood of an earthquake occurring during the life of the structure, in seismic Zones 0 and 1 (EM 1110-2-1902), no reduction of E_s or n_h is made for cyclic loading due to short-term nature of the loading. In seismic Zones 2, 3, and 4, the potential liquefaction should be evaluated. If soils in the foundation or surrounding area are subject to liquefaction, the removal or densification of the liquefiable material will be necessary. Once the designer is assured that the foundation material will not liquefy, the analysis should be performed by Saul's approach (Item 54) extended for seismic analysis as implemented in the computer program CPGD (Item 15).

(3) Stiffness Reduction Factor Equations.

(a) E_s-type Soil. For soils with a constant modulus of horizontal subgrade reaction, the following equations apply:

$$E_{s_{group}} = \frac{E_s}{R_g}$$

$$E_{s_{cyclic}} = \frac{E_s}{R_c}$$

$$E_{s_{group \ and \ cyclic}} = \frac{E_s}{(R_g R_c)}$$

$$Y_{t_{group}} = Y_t \ R_g^{0.75}$$

$$Y_{t_{cyclic}} = Y_t \ R_c^{0.75}$$

$$Y_{t_{group \ and \ cyclic}} = Y_t \ R_g^{0.75} \ R_c^{0.75}$$

(b) n_h-type Soil. For soils with a linearly increasing modulus of horizontal subgrade reaction, the following equations apply:

$$n_{h_{group}} = \frac{n_h}{R_g}$$

$$n_{h_{cyclic}} = \frac{n_h}{R_c}$$

$$n_{h_{group \ and \ cyclic}} = \frac{n_h}{(R_g R_c)}$$

$$Y_{t_{group}} = Y_t \ R_g^{0.6}$$

$$Y_{t_{cyclic}} = Y_t \ R_c^{0.6}$$

$$Y_{t_{group \ and \ cyclic}} = Y_t \ R_g^{0.6} \ R_c^{0.6}$$

(c) Definitions. The terms used in the above equations are:

$E_{s_{group}}$ = modulus of horizontal subgrade reaction for a pile in a pile group with static loading

E_s = modulus of horizontal subgrade reaction for a single pile with static loading

R_g = group reduction factor

4-36

$E_{s_{cyclic}}$ = modulus of horizontal subgrade reaction for a single pile with cyclic loading

R_c = cyclic loading reduction factor

$E_{s_{group\ and\ cyclic}}$ = modulus of horizontal subgrade reaction for a pile in a pile group with cyclic loading

$Y_{t_{group}}$ = pile head horizontal deflection at the ground surface for a pile in a pile group with static loading

Y_t = pile head horizontal deflection at the ground surface for a single pile with static loading

$Y_{t_{cyclic}}$ = pile head horizontal deflection at the ground surface for a single pile with cyclic loading

$Y_{t_{group\ and\ cyclic}}$ = pile head horizontal deflection at the ground surface for a pile in a pile group with cyclic loading

$n_{h_{group}}$ = constant of horizontal subgrade reaction for a pile in a pile group with static loading

n_h = constant of horizontal subgrade reaction for a single pile with static loading

$n_{h_{cyclic}}$ = constant of horizontal subgrade reaction for a single pile with cyclic loading

$n_{h_{group\ and\ cyclic}}$ = constant of horizontal subgrade reaction for a pile in a pile group with cyclic loading

(4) Pile Length. All of the lateral pile stiffness terms are based on the assumption that the piles are long and flexible as opposed to short and rigid. Piles are considered long if the applied lateral load at the head has no significant effect on the tip (the tip does not rotate or translate). Short piles behave rigidly and exhibit relatively no curvature (the tip rotates and translates). The computer programs referenced in this manual for group pile design are not intended for design of foundations containing short piles. Most piles used in the design of civil works structures are classified as long piles. The determination of the behavior of a pile as long or short is:

(a) Constant Modulus of Horizontal Subgrade Reaction.

$$R = \sqrt[4]{\frac{EI}{E_s}}$$

4-37

$$L/R \le 2.0 \; ; \; \text{Short pile}$$

$$2.0 < L/R < 4.0 \quad \text{Intermediate}$$

$$L/R \ge 4.0 \quad \text{Long pile}$$

(b) Linearly Increasing Modulus of Horizontal Subgrade Reaction.

$$T = \sqrt[5]{\frac{EI}{n_h}}$$

$$L/R \le 2.0 \quad \text{Short pile}$$

$$2.0 < L/R < 4.0 \quad \text{Intermediate}$$

$$L/R \ge 4.0 \quad \text{Long pile}$$

f. Torsional Stiffness. The torsional pile stiffness is expressed as:

$$b_{66} = C_{66} \frac{JG}{L}$$

where

b_{66} = torsional pile stiffness

C_{66} = constant which accounts for the interaction between the soil and the pile

J = polar moment of inertia of the pile

G = shear modulus of the pile.

L = length of the pile

The torsional stiffness of individual piles contributes little to the stiffness of a pile group for rigid pile caps and has been neglected in the past. More recent research has shown that a reasonable torsional stiffness is to use C_{66} equal to two. The coefficient C_{66} is equal to zero if the pile head is not fixed into the pile cap. See Items 44, 55, and 57 for details.

4-6. Design Procedure.

a. General. The following paragraphs outline a step-by-step procedure to design an economical pile foundation. The steps range from selection of applicable loads and design criteria through use of rigid and flexible base analyses. Identification and evaluation of foundation alternatives, including selection of the type of pile, are presented in Chapter 2.

b. Selection of Pile-Soil Model. A computer model (CPGS) is currently being developed and its capabilities are discussed in paragraph 1-3c(3), for analyzing the nonlinear interaction of the pile and surrounding soil. This model represents the lateral and axial behavior of a single pile under loading and accounts for layered soil, water table, skin friction, end bearing, and group effects. This computer model will be presented in detail in Mode CPGS. For large pile groups, the pile response is approximated by linear elastic springs. These springs represent the six degrees of freedom at the pile head, and their stiffnesses should be determined in close coordination between structural and geotechnical engineers. The designer should select a linear, elastic pile stiffness value for the group analysis by assuming a limiting deflection at the pile head. Then a secant pile stiffness should be determined for the assumed deflection using the nonlinear model or data from load tests conducted at the site. Deformations computed in the pile group analysis should be limited to this assumed deflection. The forces computed in the pile group analysis, using the secant pile stiffnesses, should be less than the actual forces from a nonlinear analysis (Figure 4-10). If more than 10 percent of the piles exceed the limiting deflection, a new secant pile stiffness should be developed for a larger limiting deflection. This method should be used in conjunction with interpretations of full-scale pile tests done at other sites that closely relate to the site under analysis. If site conditions are such that the foundation properties are not well defined, then a parametric approach should be used. A parametric analysis is performed by using stiff and weak values for the elastic springs based on predicted limits of pile group deflections. This parametric analysis should be applied to the lateral and the axial stiffnesses. See paragraph 4-5e for further discussion.

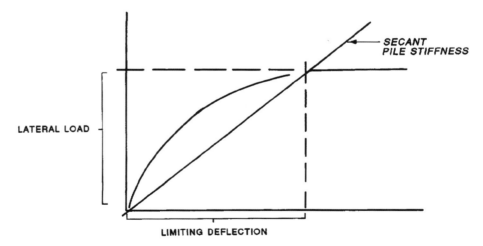

Figure 4-10. Pile forces for linear and nonlinear analysis

c. Selection of Pile Structure Model. The selection of the pile-structure model for analysis and design of a pile-founded structure must consider the following three critical items:

Type of structure (concrete or steel)

Type of analysis (rigid or flexible base)

Pile-head fixity (fixed, pinned, or partially fixed)

A reinforced concrete structure will require a rigid or flexible base analysis
with the pile heads fixed or pinned. The decision regarding which type of
base analysis to use is determined from the parametric analysis discussed in
the preceding paragraph. A rigid base analysis should use the program CPGA
(Item 5). A flexible base analysis should use one of the general purpose
finite element computer programs, such as STRUDL or SAP, which have a pile
element similar to the one used in CPGA. The flexible base analysis should be
capable of handling all degrees of freedom for the two- or three-dimensional
models. For example, to analyze a pile group with loading in the x, y, and z
directions, the base should be modeled using plate elements or three-
dimensional elements. For structures with loads in two directions only, a
typical base strip should be modeled using frame elements as shown in Figure 4-11. Pile forces and moments and structure forces and moments are
obtained from these analyses. An analysis of a steel frame on a pile foundation is accomplished in a similar manner. The degree of fixity of the pile to
the steel frame must be included in developing the pile stiffnesses. The
steel frame should be modeled as a space frame or plane frame supported by
linear elastic springs which account for the degree of pile-head fixity. Pile
forces and moments and frame forces and moments are obtained from this
analysis. Earthquake loading in seismic areas must be considered. The
program CPGD (Item 15) extends the three-dimensional rigid pile cap analysis
of CPGA (Item 5) to provide a simplified, yet realistic, approach for seismic
analysis of pile foundations. The CPGD program includes viscous damping of
the pile-soil system and response spectrum loading. The CPGD program should
be used during the seismic design process. Pile forces and moments are
obtained from this seismic analysis.

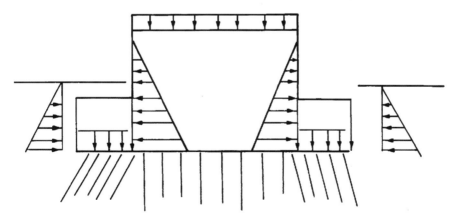

Figure 4-11. Typical 2-dimensional base strip
modeled using frame elements

    d. Selection of Load Cases. General loading conditions should be iden-
tified, and each condition should be assigned an appropriate safety factor and

allowable stress. Study of the list of loading cases will reveal that some load cases will not control the design and should be eliminated. The remaining load cases should be studied in more detail. Loading details should be established to produce critical combinations. Consider the effect each load will have on pile forces and on internal forces in the pile cap. Some loadings may control the internal design of the pile cap even though they may not produce the critical pile forces. Generally, it is important to analyze the load cases with the largest lateral loads in each direction and the cases with the maximum and minimum vertical loads. Final selection of the load cases should be based on engineering judgement.

   e. Selection of Design Criteria. Paragraph 4-2 provides specific guidance about safety factors, pile stresses, and pile cap movements. Criteria for ultimate pile capacity are presented in paragraph 4-3, and development of pile stiffness values is described in paragraph 4-5. These criteria may be applied to most pile foundation designs. However, uncertainty about pile-soil behavior may require modification of some criteria to ensure a conservative design. The magnitude of the lateral or axial pile stiffness may significantly affect the results of any pile analysis. Combining limiting values of lateral and axial pile stiffnesses may result in significantly different percentages of the applied loads being resisted by pile bending or axial force. This is particularly important for flexible base analyses because the applied loads are distributed to the piles based on the relative stiffness of the structure and the piles. Therefore realistic variations in pile stiffnesses should usually be evaluated, and the pile group should be designed for the critical condition. The variation of stiffnesses should correspond to the predicted deflection of the pile group.

   f. Deformations. The pile stiffnesses in the lateral and axial directions is determined by a nonlinear analysis assuming a limiting deformation. Since the pile stiffness is a secant model of the pile response, the calculated deflections of the pile head under working loads should be limited to that assumed value. If the analysis yields deformations greater than those assumed in determining the pile stiffnesses, then the geotechnical engineer should be consulted and the stiffnesses should be reevaluated. Calculated pile cap deformations should be checked against functional and geometric constraints on the structure. These values are usually 1/4-inch axially and 1/2-inch laterally. For unusual or extreme loads these values should be increased.

   g. Initial Layout. The simplest pile layout is one without batter piles. Such a layout should be used if the magnitude of lateral forces is small. Since all piles do not carry an equal portion of the load, axial pile capacity can be reduced to 70 percent of the computed value to provide a good starting point to determine an initial layout. In this case, the designer begins by dividing the largest vertical load on the structure by the reduced pile capacity to obtain the approximate number of piles. If there are large applied lateral forces, then batter piles are usually required. Piles with flat batters, 2.5 (V) to 1 (H), provide greater resistance to lateral loads and the less resistance to vertical loads. Piles with steep batters, 5 (V) to 1 (H), provide greater vertical resistance and less lateral resistance. The number of batter piles required to resist a given lateral load can also be estimated by assuming that the axial and lateral resistances are approximately 70 percent of computed capacity. This should be done for the steepest and flattest batters that are practical for the project, which will provide a range estimate of the number of batter piles required. For a single load case this method is not difficult. However, when the pile group is subjected to several loading conditions, some with lateral loads applied in different directions, this approach becomes more difficult. For such cases, two or three critical loading conditions should be selected to develop a preliminary layout from which the number, batters, and directions of piles are estimated. A uniform pile grid should be developed based on the estimated number of piles, the minimum pile spacing and the area of the pile cap. If piles with flat batters are located in areas of high vertical loads, then vertical piles should be placed adjacent to these battered piles. An ideal layout for flexible structures will match the pile distribution to the distribution of applied loads. This match will result in equal loads on all piles and will minimize the internal forces in the structure because the applied loads will be resisted by piles at the point of loading. For example, a U-frame lock monolith has heavy walls and a relatively thin base slab. Therefore, piles should be more closely spaced beneath the walls and located at larger spacings in the base slab.

h. Final Layout. After the preliminary layout has been developed the remaining load cases should be investigated and the pile layout revised to provide an efficient layout. The goal should be to produce a pile layout in which most piles are loaded as near to capacity as practical for the critical loading cases with tips located at the same elevation for the various pile groups within a given monolith. Adjustments to the initial layout by the addition, deletion, or relocation of piles within the layout grid system may be required. Generally, revisions to the pile batters will not be required because they were optimized during the initial pile layout. The designer is cautioned that the founding of piles at various elevations or in different strata may result in monolith instability and differential settlement.

i. Design of Pile Cap. If the pile group is analyzed with a flexible base, then the forces required to design the base are obtained directly from the structure model. If the pile group is analyzed with a rigid base, then a separate analysis is needed to determine the stresses in the pile cap. An appropriate finite element model (frame, plate and plane stress or plane strain) should be used and should include all external loads (water, concrete soil, etc.) and pile reactions. All loads should be applied as unfactored service loads. The load factors for reinforced concrete design should be applied to the resulting internal shears, moments, and thrusts acting at each cross section. The applied loads and the pile reactions should be in equilibrium. Appropriate fictitious supports may be required to provide numerical stability of some computer models. The reactions at these fictitious supports should be negligible.

4-7. Special Considerations.

a. Soil-Structure Interaction. Pile-supported structures should be analyzed based on the axial and lateral resistance of the piles alone. Additional axial or lateral resistance from contact between the base slab and the foundation material should be neglected for the following reasons. Scour of the riverbed frequently removes material from around the slab. Vibration of the structure typically causes densification of the foundation material and creates voids between the base slab and foundation material. Also, consolidation or piping of the foundation material can create voids beneath the structure.

b. Deep Seated Lateral Movement and Settlement. The soil mass surrounding a pile group must be stable without relying on the resistance of the pile foundation. In actual slides, 48-inch diameter piles have failed. Deep seated stability of the soil mass should be analyzed by neglecting the piles. Potential problems of inducing a deep seated failure due to excess pore water pressures generated during pile driving or liquefaction due to an earthquake should be recognized and accounted for in the design. The probable failure mechanism for piles penetrating a deep seated weak zone is due to formation of plastic hinges in the piles after experiencing large lateral displacements. Movement in the weak zone will induce bending in the piles as shown in Figure 4-12. A second mechanism is a shear failure of the piles which can only occur if the piles penetrate a very thin, weak zone which is confined by relatively rigid strata. The shear force on the piles can be estimated along the prescribed sliding surface shown in Figure 4-13. Research is being sponsored at the University of Texas which will develop a practical approach to solve these problems. The results of this research will be included in this manual and the capabilities of CPGA (Item 5) and CPGS, paragraph 1-3c(3), for analyzing such situations will be extended. Downdrag due to settlement of the adjacent soil mass may induce additional loads in the piles.

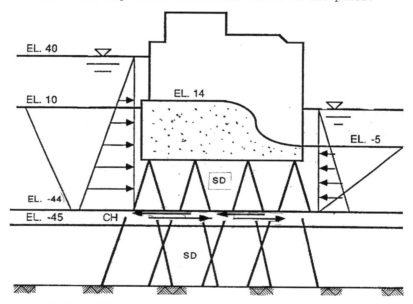

Figure 4-12. Piles are sheared off by the massive soil movement

c. Differential Loadings on Sheet-Pile Cutoffs. The length of a sheet pile cutoff should be determined from a flow net or other type of seepage analysis. The net pressure acting on the cutoff is the algebraic sum of the unbalanced earth and water pressures. The resulting shear and moment from the net pressure diagram should be applied to the structure. For flexible steel sheet piles the unbalanced load transferred to the structure may be negligible. For a continuous rigid cutoff, such as a concrete cutoff, the unbalanced load should be accounted for. An example is shown in Figure 4-14.

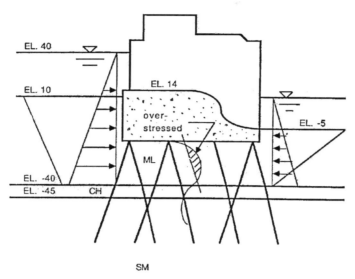

EL. 40

EL. 10

EL. 14

over-
stressed

EL. -5

ML

EL. -40

EL. -45   CH

SM

Figure 4-13.   Piles are overstressed by bending moment

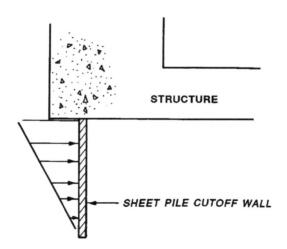

STRUCTURE

SHEET PILE CUTOFF WALL

Figure 4-14.   Pressure distribution on sheet
pile cut off wall

d. Effects of Changes in the Pile Stiffness.

(1) General. Accurate predictions of the soil-pile stiffnesses for a specific site and set of construction circumstances are extremely difficult. The interaction of the structure-pile-soil system is complex and is usually nonlinear. The load deformation behavior of this system is affected to varying degrees by the type of loading, pile spacing, pile head fixity, subgrade modulus, pile-driving procedures, water table variations, and other variables. The designer should account for these uncertainties and variations by judiciously selecting a realistic range of pile stiffnesses, and by evaluating the sensitivity of the pile forces, moments and displacements to reasonable variations in the pile stiffnesses. This procedure should be used to develop a high degree of confidence in the design.

(2) Rigid Base. For a pile group that contains only vertical piles, the rigid cap assumption requires that the plane of the pile heads remains plane when loads are applied. Therefore, since the axial and lateral components of the pile reactions are independent, changes in the axial or lateral pile stiffnesses will have predictable results. If the pile layout contains a combination of vertical and batter piles, then the interaction of lateral and axial components of the pile reactions can have significant and often unforeseen effects on the distribution of pile forces. Therefore, changes in the lateral stiffnesses could have a profound effect on the axial pile forces, and the sensitivity of the pile forces to changes in the pile stiffnesses would not be predictable without using a computer analysis. See Item 3 for example.

(3) Flexible Base. When the stiffness of the structure is not infinite compared to the stiffness of the pile-soil system, the pile cap is assumed flexible. The sensitivity of the pile loads to changes in the pile stiffness then becomes even more difficult to predict. The axial and lateral response of the piles are interrelated, and the internal stiffness of the structure significantly influences the distribution of the individual pile loads. Changes in the pile stiffnesses can also affect the deformation characteristics of the structure, thereby changing the internal moments and member forces. Figure 4-15 illustrates the effects of changing the stiffness of pile cap. In Figure 4-16 the base of the infinitely rigid pile cap deflects uniformly, causing uniform loads in the piles and large bending moments in the base slab. If the slab stiffness is modeled more realistically, as shown in Figure 4-15, the pile loads will vary with the applied load distribution. The pile loads will be lower under the base slab causing the base slab moments to be reduced. The correct stiffness relationship between the structure and the foundation is extremely important for accurately designing a pile group.

(4) Confidence Limits. An essential element of all pile foundation designs is the effort required to define the stiffness of the structure-pile-soil system confidently. Initial pile stiffnesses should be selected and used to perform a preliminary analysis of critical load cases. If the preliminary analysis indicate that the selected pile stiffnesses are not sufficiently reliable, and that the variation of the pile stiffnesses will significantly affect the analytical results, then more intensive investigation is required. Normally a limit analysis is performed to bracket the solution. With this limit approach, all the factors which tend to minimize the pile-soil resistance are collectively used to represent a weak set of pile stiffnesses. This condition is a lower bound. Similarly, all the factors which tend to maximize

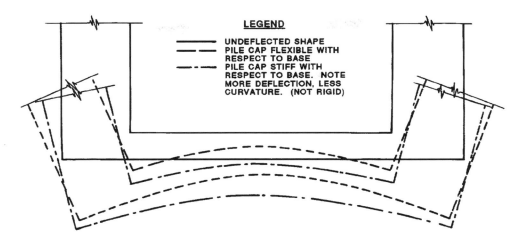

Figure 4-15. Deflected shape of flexible pile caps

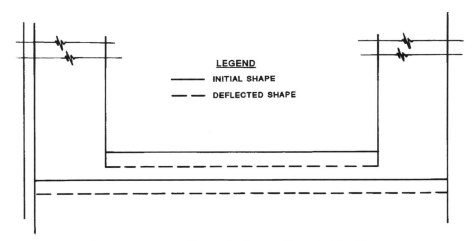

Figure 4-16. Deflected shape of a rigid pile cap

pile-soil resistance are collectively used to represent a strong set of pile stiffnesses (upper bound). Using this procedure, the designer can establish confidence limits by performing two analyses which bracket the actual set of parameters. For further discussion of this procedure, refer to Paragraph 4-6.

e. Effects of Adjacent Structures.

(1) General. Most hydraulic structures are designed to function as independent monoliths. Sometimes it is necessary to design hydraulic structures which interact with adjacent monoliths or existing structures. Certain procedures and details should be used to assure that the actual structural performance is consistent with the design assumptions.

(2) Independent Monoliths. Generally, hydraulic structures should be designed to function as independent monoliths. Each monolith should be isolated by vertical joints and should not interact with adjacent monoliths. This approach greatly simplifies the analysis and is a reliable basis for predicting performance. Validity of the design assumptions should be assured by including the following procedures and details. Independent monoliths should not be physically connected to adjacent monoliths. Expansion joints should be provided between monoliths to accommodate the predicted displacements. Rigid cap displacements should be extrapolated to the top of the monolith and the displaced structure should not make contact with adjacent monoliths. Batter piles should not interfere (based on common construction tolerances) with piles under adjacent structures. It is good design practice, but not always practical, to keep the tips of all piles within the perimeter of the pile cap. Possible interferences with piles under adjacent monoliths should be checked using CPGI, a program currently being developed and discussed in paragraph 1-3c(6), and the pile layout should be modified as needed.

(3) Interacting Monoliths. Sometimes it is necessary to design the pile groups of adjacent monoliths to interact and resist large unbalanced lateral loads. There are three types of circumstances:

(a) Analysis of new structures that are geometrically constrained from permitting sufficient batter and numbers of vertical piles to resist the lateral forces.

(b) Analysis of new structures that are subjected to a highly improbable loading condition. Such extreme lateral loads make it economically unfeasible to design a pile layout for independent adjacent monoliths.

(c) Evaluation of existing structures.

For designing new structures, provisions should be included to assure positive load transfer between monoliths (preferably at the pile cap) and without causing detrimental cracking or spalling. Provisions should be included for keying and grouting the monolith joint between the pile caps of interacting structures. Special attention should be given to the monolith joints in the thin wall stems of U-frame locks. The wall joints should be detailed to accommodate monolith movements without significant load transfer and thereby control localized cracking and spalling. For evaluating existing structures, the analyst should model actual field conditions as closely as practical. Field measurements should be made to determine pile-cap displacements and changes in monolith joint dimensions. Investigations should determine if load transfer is occurring between monoliths (joint closure, spalling concrete at joints, etc.). Foundation investigations should be adequate to estimate the lateral and axial pile stiffnesses.

(4) New Structures Adjacent to Existing Structures. Special provisions are appropriate for designing and installing piles adjacent to an existing structure. Existing structures include those under construction or already in service. During construction, pile driving should not be allowed within 100 feet of concrete which has not attained its design strength. Pile driving within 100 feet of concrete that has achieved the required design strength should be monitored for detrimental effects on the existing concrete. If piles are installed near an existing structure, it is prudent to monitor and document effects of pile driving on the existing structure and foundation.

Such provisions should be fully considered during design. Potential damage to existing structures may be influenced by a variety of factors:

(a) Densification of existing fill may induce settlement and a significant increase in lateral earth pressures.

(b) Driving displacement piles in noncompressible materials may cause heave of the ground surface.

(c) Driving piles in submerged, uniformly fine-grained, cohesionless soils may rearrange the soil grains and increase groundwater pressure with corresponding large settlements.

(d) Lateral load resistance of adjacent pile foundations may be significantly reduced.

These factors and others should be thoroughly investigated during design.

(5) Special Techniques. Special types of pile installation should be used to minimize possible damage. These may include:

(a) Using nondisplacement piles.

(b) Specifying a pile hammer that minimizes vibrations.

(c) Jacking piles.

(d) Using predrilled pilot holes or jetting.

The condition of existing structures and the surrounding area should be carefully documented before, during, and after pile driving. Field surveys, measurements, photographs, observations, sketches, etc. should be filed for future reference.

f. Overstressed Piles. The design criteria in preceding paragraphs are generally applicable for each load case. However, on large foundations, a few piles may exceed the allowable capacity or stresses by a relatively small amount without endangering the integrity of the structure. The design of a pile group should not be dictated by localized overload of a few corner piles for one load case. Because of the highly nonlinear load-deflection relationship of piles and the large plastic ranges that some piles exhibit, high localized pile loads are usually redistributed without danger of distress to adjacent piles until a stable state of equilibrium is attained. The stiffness method of analysis is an approximate linear model of the nonlinear load deflection behavior of each pile. Since the stiffness analysis is not exact a few piles may be loaded above the allowable capacity. Iterative pile group analyses are required.

g. Pile Buckling. Buckling of individual piles is related to the load level, the flexibility of the pile cap, the geometry of the pile group, and the properties of the soil and piles. Pile-soil stiffness and the degree of lateral support provided by the soil primarily depend on the following factors:

(1) Embedment. If the piles are fully embedded, then the lateral support provided by the soil is usually sufficient to prevent pile buckling. Even extremely weak soils may provide sufficient support to prevent buckling when fully embedded. Buckling may be critical if the piles project above the surface of soils that provide strong lateral support.

(2) Rigidity. The pile shape (radius of gyration), modulus of elasticity of the pile, the lateral and axial support provided by the soil, the degree of fixity of the pile head, and the flexibility of the pile cap all affect the relative pile rigidity. Buckling analysis is very complex because the axial and transverse loads and the pile stiffnesses affect the deformation of the pile, and this behavior is related through interaction with the soil.

(3) Tip Resistance.

h. Pile Splicing.

(1) General. The probability and reliability of splicing piles should be considered early in design. The structural integrity of the piles and complexity of the installation procedures must be comprehensively evaluated before selecting the location and types of splices allowed. Most splicing is performed in the field and significantly increases construction time, cost, and the field inspection required to assure reliability. Therefore, field splicing is normally limited to situations where only occasional splices are required. Splicing may be necessary in construction areas with limited overhead clearances or if the pile does not attain its required design capacity at the specified tip elevation. Contract plans and specifications should address the use (or exclusion) of splices and any specific requirements or limitations that are necessary. Splicing should not be allowed in the field without prior consent and approval of the designer.

(2) Structural Integrity. Splices should be capable of resisting all forces, stresses, and deformations associated with handling, driving, service loads, or other probable sources. Splices in the upper portion of the pile should be designed to account for the possible effects of accidental eccentric loadings. Regions of low bending and shear stresses under service loads are preferable for splice locations. Allowable stresses should be limited to those listed in paragraph 4-2d, and deformations should be compatible with the interaction between the pile and structure. The design should also account for the effects of corrosion and cyclic or reverse loading if present. Many commercial splices are not capable of developing the full strength of the pile in tension, shear, and bending.

(3) Soil Integrity. Splice surfaces which extend beyond the perimeter of the pile may disturb the interface between the pile and soil during driving and decrease adhesion. If appropriate, reductions in axial and lateral pile capacities should be made. This condition is most likely to occur in stiff clays, shales, and permafrost.

(4) Installation. Most splicing is performed in the field, sometimes in the driving leads. Engineering experience and judgement are essential in assessing the critical factors influencing reliability and cost (i.e. field access to the splice location, workmanship and quality assurance). The time required to perform the splice is also critical if the pile tends to set and become more difficult to restart when driving resumes. Piles driven into

materials with high adhesion or granular materials exhibit rapid set to a greater degree than soft clays or sensitive soils.

i. As-Built Analyses. As explained in paragraph 5-6a, conditions encountered in the field may result in variations between the pile foundation design and the actual pile foundation. All such variations should be observed, recorded and evaluated by the designer in an as-built analysis. The number of overloaded piles, the severity of the overload, and the consequences of the failing of one or more overloaded piles should be evaluated in the as-built analysis. Structural deformations and interaction between adjacent monoliths also could be significant factors.

(1) Geometric Factors. Field conditions may cause variations in the geometric layout of individual piles; i.e., pile head may move horizontally or rotate, batter may change, and final tip elevation may vary due to a change in batter or soil properties. Such geometric variations may substantially affect the individual pile loads even though the pile capacity remains unchanged.

(2) Soil Properties. Variations in soil properties may affect pile-tip elevations, pile capacities and the axial and lateral pile stiffnesses.

(3) Obstructions. Unexpected subsurface obstructions may prevent driving some piles to the design tip elevation, thereby causing variations in the pile stiffnesses or necessitating field changes.

CHAPTER 5

ENGINEERING CONSIDERATIONS PERTAINING TO CONSTRUCTION

5-1. <u>General</u>. This chapter addresses engineering considerations pertaining to the construction of pile foundations. It is important for the designer to become familiar with the various equipment (Items 31 and 32) and methods used during construction since either may adversely affect soil-structure interaction, economics, and the overall effectiveness of the design. Early in the design process consideration should be given to available pile materials and lengths, appropriate construction methods and equipment, load tests, acceptable and achievable construction tolerances, and maintaining quality control and records during construction. Design coordination with construction should begin in the early design stages. These considerations, combined with past experience, should result in the formulation of an appropriate foundation design and the preparation of suitable construction plans and specifications. Upon completion, a review of construction variations should be made to determine if an as-built analysis is warranted. Material presented in this chapter is intended to give design and construction engineers an overview of installation and its effect on the design. Detailed discussions can be found in the literature and the cited references.

5-2. <u>Construction Practices and Equipment</u>. A variety of methods and special equipment have been used for the installation of piles. Many factors are involved in the selection process, but the end result should always be structurally sound piling supported by soil or rock that is capable of developing the design loads. To achieve this result, it is imperative that the specifications provide for use of appropriate installation methods, suitable equipment, and acceptable field procedures. Contract specifications should be as flexible as possible so that economy is achieved, yet rigid enough to result in the desired final product.

a. Installation Practices. Installation practices include consideration and utilization of appropriate field methods for storing, handling, and accurately driving each pile to the desired final position within established tolerances. Specifications typically outline requirements for the contractor to submit his proposed plan for installing the pile. Required submittal items normally include detailed descriptions for pile storage and handling, the driving rig and all auxiliary equipment (with manufacturer's specifications and ratings), installation sequence, methods for controlling placement and alignment of piles, and, if permitted, the pile splice types, locations and plan, and quality control plan. In addition, the specifications normally require submittal of data for a Government-performed wave equation analysis. Government review should focus on the contractor's compliance with the specifications and the ability of his proposed equipment and methods to produce structurally sound piling, driven within the established tolerances and capable of developing the required design capacity. Installation methods or equipment suspected of compromising the foundation design should be clearly excluded by the specifications. The contractor may question those exclusions and may substantiate his claim at his expense by performing wave equation analysis, field verification of driving and static load tests, dynamic monitoring, or other methods designated by the designer.

(1) Storage and Handling. Piles are subject to structural damage during the storage and handling processes. Improper storage or handling may result

5-1

in excessive sweep (camber) or cracking in concrete and may be cause for rejection of a pile. Excessive sweep, or camber, has been known to result in a pile drifting out of tolerance during installation. Sweep and camber limitations should be included in the specifications. Stresses developed during the storage and handling phases should be investigated and compared to those allowed in paragraph 4-2d. Additionally, both the required number and locations of permissible pick-up points on the pile should be clearly indicated in the plans and specifications. Any deviations in the field must be approved by the design engineer. Special care must be exercised when handling piles with protective coatings, and damaged areas must be repaired prior to installation. All pilings should be visually examined at the driving site by a qualified inspector to prevent the use of any pile damaged by faulty storage or handling procedures.

(2) Placement and Tolerances. When determining suitable placement tolerances, consideration should be given to the site conditions, i.e., topography; subsurface materials; type of loading; pile type, spacing and batter; size and type of pile cap and structure; available driving equipment; and possible interference between piles. A lateral deviation from the specified location at the head of not more than 3 to 6 inches measured horizontally and a final variation in alignment of not more than 0.25 inch per foot measured along the longitudinal axis should normally be permitted. In the vertical direction a deviation of plus or minus 1 inch from the specified cutoff elevation can be considered reasonable. The above recommendations are general guidance for large pile groups and should be verified as applicable for each specific project. It should be noted that sloping surfaces may require field adjustment of the pile location if the actual excavation line differs from the reference plane used in the plans to depict pile locations. Each pile should be checked in the field prior to driving. The pile head should be seated in the hammer and the pile checked for correct batter, vertical plumbness, and rotation of the pile by a method approved by the design engineer. Many jobs require the use of a transit to set the pile and the leads accurately when driving battered piles. Once driving has commenced, attempts to move the pile by rotating the leads, pulling on the pile, or wedging may result in damage (structural or soil alteration) or increased misalignment of the pile tip.

(3) Driving. Contract specifications disallow field driving of piles until the contractor's methods and equipment are approved by the design engineer. Designer approval is necessary to ensure the pile can be driven without damage to the pile or soil, and methods for determining such are discussed in paragraph 5-3. The designer should be aware that certain equipment and methods for pile installation have been known to reduce axial and lateral resistance or damage the pile in certain situations. Field variations from the approved methods and equipment require re-submittal to the design office, as changes can and usually do effect the pile capacity attained for a given length pile. It is incumbent upon the designer to supply field personnel with the necessary information to ensure each pile installed is capable of supporting its design load. Such information most often consists of limiting penetration resistances (paragraph 5-3) or the specification of a pile driving analyzer (paragraph 5-4a) to prevent structural damage from overdriving and to ensure that adequate capacity is developed. Field personnel must ascertain the equipment and installation methods are properly employed, the equipment is performing up to its capabilities, records are properly kept (paragraph 5-4b), and any driving abnormalities are promptly reported back to the design office. Pile driving should not result in crushing or spalling of concrete, permanent

deformation of steel, or splitting or brooming of wood. Damage sustained
during driving can frequently be attributed to misalignment of the pile and
hammer, a material failure within the drive cap, equipment malfunction, or
other improper construction practices. Field installation requires diligent
monitoring of penetration resistance. Any piling suspected of either sustaining structural damage or failing to develop the required capacity, for
whatever reason, must be promptly evaluated by the designer to determine its
effect on the overall foundation design. Repetitive problems may require
modification of the installation equipment or procedure. Pile heave can be a
problem in some cases and is more inclined to occur for displacement piles.
In this case, an installation sequence should be required to minimize the
likelihood of pile heave. Piles that experience heave should be restruck to
seat the pile properly. The installation of a concrete pile requires special
consideration due to its inherent low tensile strength. The pile must be
firmly seated prior to the application of full driving energy to prevent pile
cracking or breakage. Pile driving can sometimes be supplemented by special
driving assistance such as the addition of driving shoes, jetting, preboring,
spudding, or followers. The use of special assistance should be considered
when one of two conditions exist. If a pile reaches refusal with a suitable
hammer but does not achieve the necessary capacity, a modification to the
installation procedures may be necessary. Simply increasing the size of the
hammer may not be appropriate because the pile would be damaged due to
excessive driving stresses. The second condition is an economic one, where
the installation time and effort can be substantially reduced by the modifying
installation procedures. In either case, the potential effect on the axial
and lateral pile capacity must be closely evaluated. Contract specifications
should define as clearly as possible what type of special driving assistance,
if any, would be allowed and under what conditions they would be allowed.
Since methods of providing special driving assistance usually result in
reduced pile capacity, specifications normally preclude their use without
written approval from the designer. Methods and rationale for the selection
of equipment, field inspection, establishment of penetration limitations,
record keeping requirements and methods for controlling the driving operation
are contained elsewhere in this chapter.

(a) Pile shoes. Pile shoes are frequently used to improve driveability
and also provide protection at the pile tip. When driving piles in dense
sands, in hard layers containing cobbles or boulders, or through other
obstructions, increased cutting ability and tip protection are provided by the
shoe. Piles seated in rock normally require shoes for tip protection and
improved bearing characteristics. Steel pile shoes are usually fabricated of
cast steel, particularly for steel H-piles, where plates welded to the flange
and web have proven unreliable. The design engineer should evaluate the
necessity and cost of using pile shoes on a case-by-case basis.

(b) Jetting. Jetting is normally used when displacement-type piles are
required to penetrate strata of dense, cohesionless soils. Exceptions are
very coarse or loose gravel where experience shows jetting to be ineffective.
Piles, in some cases, have been successfully jetted in cohesive soils but clay
particles tend to plug the jets. Jetting aids in preventing structural damage
to the pile from overdriving. Water is pumped under high pressure through
pipes internally or externally attached to the pile, although air may be used
in combination with the water to increase the effectiveness in certain cases.
The last 5 to 10 feet of pile penetration should be accomplished with no
jetting allowed. Piles that cannot be driven the last 5 to 10 feet without

the aid of jetting should be immediately brought to the attention of the design engineer, since a reduction in axial capacity will probably result. When jetting concrete piles, driving should be restricted to a static weight while the water is being injected to prevent damage due to excessive tensile stresses that may be induced by impact. Jetting adjacent to existing structures or piles should be avoided if possible. Although driving vibrations are reduced, extreme caution must be exercised, since jetting causes disturbance of soil material. The design engineer must exercise caution when determining the design capacity for a jetted pile. Adequate provisions must be made for the control, treatment (if necessary), and disposal of run-off water. If jetting is anticipated, test piles should be installed using jetting, with the test pile being installed after the reaction piles are installed to assess the effects of jetting on capacity.

(c) Preboring. A pilot or prebore hole may be required to penetrate hard nonbearing strata; to maintain accurate location and alignment when passing through materials which tend to deflect the pile; to avoid possible damage to adjacent structures by reducing vibrations; to prevent heave of adjacent buildings; or to remove a specified amount of soil when installing displacement-type piles, thereby reducing foundation heave. Preboring normally takes place in cohesive soils and is usually required when concrete piles must penetrate man-made fills and embankments containing rock particles or other obstructions. It should be noted that on past Corps projects, concrete piles have been successfully driven through man-made fills such as levee embankments without preboring. Preboring through cohesionless soils is not recommended, since the prebored hole may not stay open and could require a casing. The most widely used method of preboring is by utilizing an auger attached to the side of the crane leads. When preboring is permitted, the hole diameter should not be greater than two-thirds the diameter or width of the pile and not extend more than three-fourths the length of the pile. Oversizing the hole will result in a loss of skin friction and a reduction in the axial capacity and lateral support, thereby necessitating reevaluation of the pile foundation. When extensive preboring is needed, consideration should be given to using a drilled-shaft system rather than a driven-pile system.

(d) Spudding. Spudding is similar to preboring and may be appropriate when layers or obstructions are present near the surface that would damage the pile or present unusual driving difficulty. Spudding is accomplished by driving a spud, such as mandrel, heavy steel pipe or H-pile section, to provide a pilot hole. The spud is withdrawn and the pile inserted into the hole and driven to the required depth. Problems may result if the spud is driven too deep, since extraction becomes more difficult as penetration is increased. Spudding may sometimes entail alternately lifting a partially driven pile a short distance and redriving it when very difficult driving is encountered (e.g. for heavy piles). Because this procedure adversely affects the soil's lateral and axial capacity, it should be avoided for friction piles and should never be permitted without the specific authorization of the design engineer.

(e) Followers. A follower is a member placed between the pile hammer and pile that allows the pile to be driven below the reach of the leads. The most common uses are to drive a pile below the top of an existing structure or for driving piles over water. Although the follower can make driving less difficult, there are several problems associated with their use. Experience shows it to be quite difficult to maintain alignment between the pile and follower, especially for battered piles. Additionally, erratic energy losses due to poor connection between the pile and follower, frequent misalignment, and follower flexibility make it nearly impossible to equate blow count with pile capacity. For these reasons most specifications exclude the use of followers. If a follower must be used, it should be selected so that it's impedance is between 50 and 200 percent of the pile impedance. The impedance is defined as EA/c where E is the modulus of elasticity of the material, A is the cross sectional area, and c is the velocity of wave propagation for the material. If concrete piles are being driven, then some cushion must be used between the follower and the pile.

(4) Extraction. Extraction, or pulling of specific piles for inspection, may be required when unusually difficult driving conditions have been encountered and pile damage is suspected. Extraction and redriving may also be necessary when a pile drifts excessively during driving and fails to maintain the specified placement tolerances discussed in paragraph 5-2a(2). When excessive drift occurs, the circumstances should be carefully investigated to determine the cause and appropriate remedial measures specified. Pile extraction is usually difficult, expensive, requires special equipment and experienced personnel. A large pulling force concentric with the longitudinal axis of the pile must be exerted continuously in addition to application of hammer energy in the same direction. Extraction can be assisted by jetting, especially when removing low tensile strength piles such as concrete. See paragraph 5-2b(2) for a discussion of equipment required for extraction.

(5) Underwater Driving. Occasionally piles must be driven below the water surface at a location where site dewatering is not economically feasible, e.g., navigation fenders, dolphins, guide walls, piers, etc. Commonly, pile driving equipment is placed on barges and positioned at the work site with tug boats. A special templet is normally utilized to maintain the designated position and alignment of the piles during driving. Placement tolerances are usually less stringent for these structures. When the pile head must be driven below the water surface, a follower with a special connection to the pile head may be used. In some cases a hydraulically driven, submersible pile hammer (clamped to the pile head) may be used, especially if the pile head must be driven a substantial distance below the water surface. For example, a submersible hammer would be appropriate to drive steel piles to the mudline for anchoring mooring buoys that have substantial design loads and the accuracy of placement position is not critical.

b. Equipment. Piles are normally driven by impact or vibratory-type hammers. Typical driving equipment consists of a crawler-mounted crane with a boom, leads, hammer, and various accessories, each connected to act as a unit. The equipment serves to guide and drive each pile accurately into its final position and must be strong enough to withstand safely all loads imposed during the process. The crane and boom must have adequate size, capacity and connections to handle the pile and the special driving equipment, such as the hammer or extractor, leads, and accessories, safely. Considerable engineering experience and judgement are necessary when evaluating or specifying the suitability of driving equipment. Supplemental information is normally available in the form of technical literature provided by the equipment manufacturer. Only equipment anticipated to be detrimental to the pile, soil, or soil-pile interaction should be excluded by the construction specifications. A discussion of hammer selection is presented in paragraph 5-3b. Safe equipment operation must also be considered in the design and construction phases of a project. Common situations that typically require special safety precautions are obstructions (such as overhead or buried electrical lines), driving on slopes or near the edges of excavations, and possible crane overturning. Specific safety requirements are contained in EM 385-1-1.

(1) Hammers. Hammers can generally be divided into two groups, impact and vibratory. Impact hammers may be lifted manually or automatically by steam, air or diesel, and may also be single or double-acting. These hammers are sized by the maximum "rated energy" (foot-pounds) theoretically contained as kinetic energy in the ram just before impact. This rated energy is not necessarily absorbed by the pile. Vibratory hammers are electrically or hydraulically powered, usually have a variable operating frequency range (vibrations per minute), and are generally rated by "eccentric moment" (inch-pounds) and "driving force" (tons) for a specified frequency. Literature providing specific properties for currently available hammers may be obtained on request from the hammer manufacturer or distributor. The hammer approved for use should be examined in the field to assure that the hammer is in good condition and operating as close as possible to its rated capacity in accordance with procedures provided by the manufacturer. Hammer efficiency may be influenced by items such as the operating pressure, wear of moving parts, lubrications, drive cap cushions, driving resistance, batter angle, and the relative weights of the hammer and pile. Operating pressure at the hammer (for steam and air hammers), stroke distance and operation rate (blows per minute) must be checked regularly while driving piles with any type of impact hammer. Variations in these values usually signify changes in hammer energy and efficiency, or pile damage. Steam- or air-powered automatic-type hammers also require special supplemental equipment, including adequately sized hoses, power source and fuel, and self-powered air compressor or boiler with a water supply for steam. A brief description of the various hammers and general recommendations follow. Item 31 contains an excellent discussion of hammer operation and suggested inspection techniques.

(a) Drop Hammers. The drop hammer is the simplest and oldest type of impact hammer. It consists of a guided weight (ram) that is lifted to a specified height (stroke) by a hoist line and released. Drop hammers are operated by raising the ram with the crane and then, at the desired height as judged by the crane operator, dropping the ram by allowing the winch to spool. Some of the available energy is used as kinetic energy in the winch and is not actually available to drive the pile. Drop hammers can damage the pile head if driving stresses are not controlled by limiting the stroke distance and supplying a cushion material (hammer cushion) between the anvil, which sits on the pile head, and ram. Theoretical or rated hammer energy is the product of the stroke times the ram weight. To arrive at actual energy delivered to the pile, proper allowances must be made for the effects of friction and interaction of the drive cap. The drop hammer is a comparatively simple device that is easily maintained, portable, relatively light, and does not require a boiler or air compressor. The drop hammer is most suitable for very small projects that require relatively small, lightweight timber, steel, or aluminum piles. Due to its slow operating rate, usually 5 to 10 blows per minute, this type of hammer is used only when the cost of bringing in a more sophisticated hammer would not be economical.

(b) Single-Acting Steam or Air Hammers. The single-acting hammer as shown in Figure 5-1 has been in use for many years, has been extremely well developed and can be used for most any pile-soil combination. This hammer type utilizes pressure from steam or compressed air to raise the ram, then

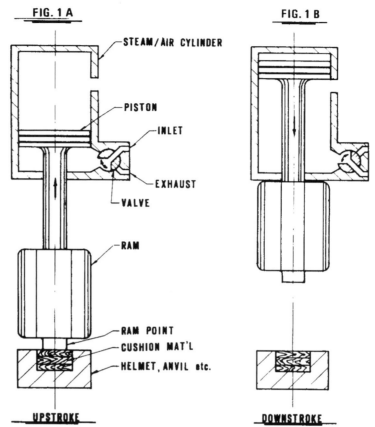

FIG. 1 A   FIG. 1 B

- STEAM/AIR CYLINDER
- PISTON
- INLET
- EXHAUST
- VALVE
- RAM
- RAM POINT
- CUSHION MAT'L
- HELMET, ANVIL etc.

UPSTROKE   DOWNSTROKE

Figure 5-1. Single-acting steam/air hammer (Permission to reprint granted by Deep Foundation Institute (Item 31))

automatically releases the pressure allowing the ram to fall freely and strike the drive cap. Hammer operation is automatic and generally in the range of 40 to 60 blows per minute. In comparison to the drop hammer, single-acting hammers operate at much faster speeds, have shorter stroke distances and possess considerably larger ram weights. A hammer cushion may or may not be utilized within the drive cap, and its use is largely dependent on the recommendations of the hammer manufacturer. Hammer efficiency can be checked by observation of the ram stroke and hammer operation rate. If the hammer maintains the specified stroke and operating speed, it can be reasonably assumed the hammer is functioning properly. A single-acting hammer may lose considerable driving energy when used to drive battered piles. This energy loss can be attributed to a reduction in the height of the ram's vertical fall and increased friction between the piston and cylinder wall and between the ram and the columns.

(c) Double-Acting Steam or Air Hammers. Double-acting and differential-acting hammers, as shown in Figures 5-2 and 5-3, utilize pressure from steam or compressed air to raise the ram in a manner similar to a single-acting

5-7

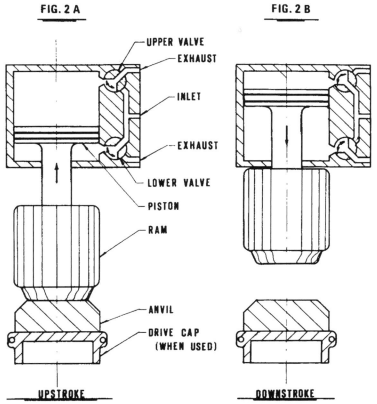

FIG. 2 A   FIG. 2 B

UPPER VALVE
EXHAUST
INLET
EXHAUST
LOWER VALVE
PISTON
RAM
ANVIL
DRIVE CAP
(WHEN USED)

UPSTROKE   DOWNSTROKE

Figure 5-2. Double-acting steam/air hammer (Permission to
reprint granted by Deep Foundations Institute (Item 31))

hammer. The steam or compressed air is also utilized to supply additional
energy to the ram on the downward part of the stroke. The combination of
pressure on the downstroke and a short stroke distance results in an operating
rate generally ranging from 90 to 150 blows per minute. These hammers can
deliver impact energies comparable to the single-acting hammers at approximately 1.5 to 2.0 times the operating rate. Although the high operation speed
is beneficial to production, it generates relatively high impact velocities
and stresses, which may result in pile-head damage to piles of low compressive
strength. A hammer cushion material is not used between the ram and pile
helmet for the double-acting hammer but is required for the differential-
acting hammer. The types of impact hammers are normally closed at the top,
and the stroke cannot be monitored during driving. Actual field operation
should be at the full hammer speed as listed by the manufacturer, since the
rated hammer energy quickly reduces at lesser speeds. Rated energy and
efficiency values provided by the manufacturers can be misleading, and the
engineer must be cautious and use appropriate judgement when calculating the
energy actually transferred to the pile during driving. These hammer types
may be used without leads (when not required for piles) and may be inverted
and rigged for use as pile extractors. Best performance is usually obtained

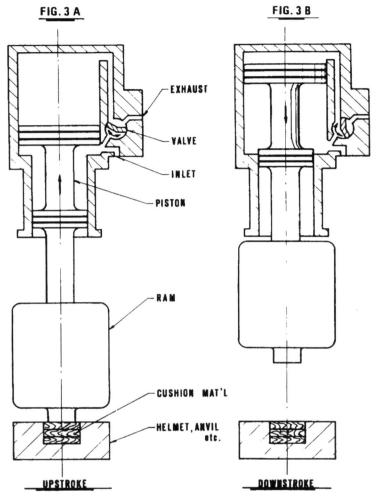

FIG. 3 A

FIG. 3 B

EXHAUST

VALVE

INLET

PISTON

RAM

CUSHION MAT'L

HELMET, ANVIL
etc.

UPSTROKE

DOWNSTROKE

Figure 5-3. Differential-acting steam/air hammer (Permission to
reprint granted by Deep Foundations Institute (Item 31))

when driving wood or nondisplacement steel piles into sands, but the hammers
may be used in any type soil.

(d) Open-End Diesel Hammers. The open-end diesel hammer (Figure 5-4),
also known as the single-acting diesel hammer, is self-contained, economical,
light in weight, and easy to service. The fuel is injected into the cylinder
while the ram drops. When the ram strikes the anvil the fuel is atomized and
ignited, explodes and forces the anvil down against the pile and the ram up.
This supplies energy to the pile in addition to that induced by impact of the
ram. The sequence repeats itself automatically provided that sufficient pile
resistance is present. Hammer efficiency is a function of pile resistance and
therefore the harder the driving the greater the efficiency. Diesel hammers
can be equipped to permit the amount of fuel injected into the cylinder to be

5-9

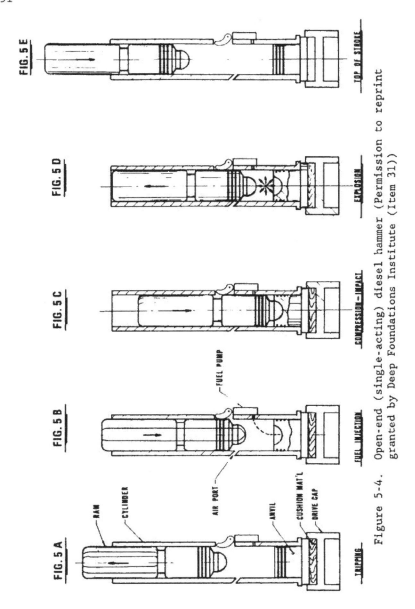

Figure 5-4. Open-end (single-acting) diesel hammer (Permission to reprint granted by Deep Foundations Institute (Item 31))

varied. This feature can be an asset when initially seating concrete pile.
The energy transmitted to the pile can be controlled by limiting the amount of
fuel supplied to the hammer, thereby yielding some control on the critical
tensile stresses induced by driving. Diesel hammers combine medium ram
weights and high impact velocities. The open-end diesel hammer requires a
cushion material (hammer cushion) between the anvil and the helmet. Operating
speeds are somewhat slower than the single-acting air-stem hammer ranging from
40 to 50 blows per minute. As the driving resistance increases, the stroke
increases and the operating speed decreases. Proper maintenance and operation
of the diesel hammer is a necessity. Open-end diesel hammers are best suited
for medium to hard driving conditions. They do not tend to operate well in
soft soils because of the driving resistance required for compression and
ignition.

    (e) Closed-End Diesel Hammers. The closed-end diesel hammer,
Figure 5-5, also known as the double-acting diesel hammer, is similar to the
open-end hammer, except that a closed top and bounce chamber (air tank) are
provided at the upper end of the cylinder. The stroke is shortened from that
of the open-end hammer by creating a cushion of compressed air in the bounce
chamber and between the ram and the closed upper end of the cylinder. This
results in operating speeds of about 80 blows per minute. Some closed-end
hammers are convertible to the single-acting mode, thereby giving the contractor further flexibility. Requirements for cushion materials, leads, maintenance, and operation are similar to those of the open-end diesel hammer.

    (f) Vibratory Hammers. Vibratory hammers are available in high, medium,
and low frequency ranges. High-frequency hammers are commonly known as "sonic
hammers." The sonic hammer has had limited success and is seldom used. Vibratory hammers operate by utilizing electric or hydraulic motors to rotate
eccentric weights and produce vertical vibrations as shown in Figure 5-6. The
vibrations reduce frictional grip of the soil and also permit the soil at the
tip to be displaced. Additional biased static loads can often be provided by
dead weight to enhance drivability. Leads are not required for use of a
vibratory hammer but are normally required for desired driving accuracy. It
is important that a rigid connection be maintained between the hammer and the
pile, usually by means of a mechanical clamp, and a back-up system may be
required to prevent release of the clamp in the event of a power failure.
Vibratory hammers are most efficient for installing non-displacement type
piles in sand. Clay soils tend to dampen the vibration of the hammer, thereby
retarding penetration. When used in clay materials, the low frequency hammer
has been more successful since it has more of a chopping effect than the
medium-frequency hammer which is normally used for sands. These hammers are
not very effective in penetrating obstacles, large cobbles or stiff clays.
Vibratory hammers are generally not suitable for the installation of most
concrete piles and are seldom used on timber piles. When used for the right
combination of pile and soil, vibratory hammers can install production piles
at a rate much faster than any type of impact hammer. For example, it would
not be uncommon to drive a 60-foot steel H-pile in sand in less than
5 minutes. An added advantage of the vibratory hammer is that it can extract
piles as easily as it can drive them, requiring no new equipment set-up.
Vibratory hammers and their limitations are discussed in paragraph 5-3b.

    (2) Extractors. The extraction of piles can be difficult and usually
requires special equipment and experienced personnel. Extractors can be classified as either impact or vibratory type. The impact type operates similar

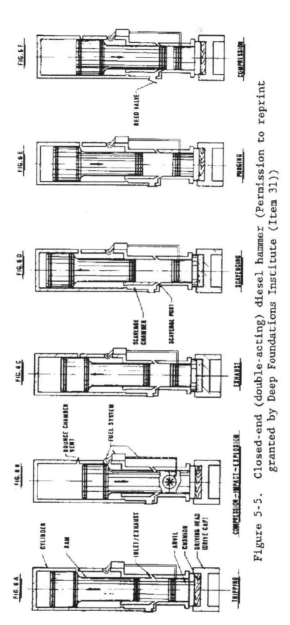

Figure 5-5. Closed-end (double-acting) diesel hammer (Permission to reprint granted by Deep Foundations Institute (Item 31))

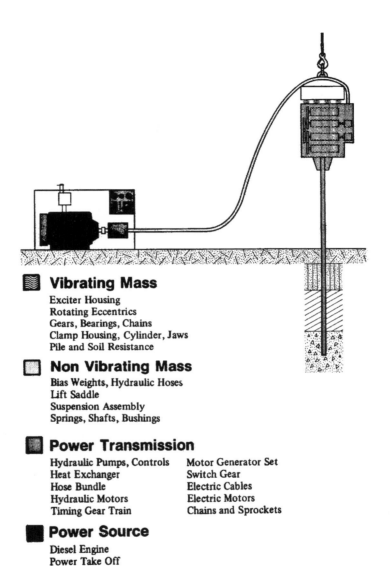

Vibrating Mass
Exciter Housing
Rotating Eccentrics
Gears, Bearings, Chains
Clamp Housing, Cylinder, Jaws
Pile and Soil Resistance

Non Vibrating Mass
Bias Weights, Hydraulic Hoses
Lift Saddle
Suspension Assembly
Springs, Shafts, Bushings

Power Transmission
Hydraulic Pumps, Controls   Motor Generator Set
Heat Exchanger   Switch Gear
Hose Bundle   Electric Cables
Hydraulic Motors   Electric Motors
Timing Gear Train   Chains and Sprockets

Power Source
Diesel Engine
Power Take Off

Figure 5-6. Vibratory driver/extractor system (Permission to
reprint granted by Deep Foundations Institute (Item 31))

to the double-acting hammer in an inverted position and is powered by compressed air or steam. The vibratory type is a vibratory hammer which is used
for extraction, is operated in the same manner as for driving except a steady
pulling force is provided and can be as effective as driving. When pulling a
pile with either type of extractor, a steady pull must be exerted through the
crane line on the pile in the direction of its longitudinal axis to supplement
the extractor energy. The lifting line of the crane is attached to the
extractor, and the extractor is connected to the pile head with rigid side

straps or clamps. This connection must be strong enough to transfer to the pile safely the large forces that are developed by the combined action of the lifting line and the extractor during the pulling operation. If the pile is vertical, or nearly vertical, leads are normally not required for extraction. However, a steeply battered pile would normally require leads to maintain the alignment of the pulling forces along the longitudinal axis of the pile. Effectiveness of the extraction process is directly related to the steady pull exerted by the crane line in the direction of the pile axis plus the efficiency of the extractor. Large hydraulic jacks have occasionally been used to jack piles out of the ground slowly under unusual circumstances, but this method of extraction is not recommended due to the excessive time required and large reaction forces generated.

(3) Leads. Pile driving leads, sometimes called leaders, are usually fabricated of steel and function to align the pile head and hammer concentrically, maintain proper pile position and alignment continuously during the driving operation, and also to provide lateral support for the pile when required. Typical lead systems are shown in Figures 5-7 and 5-8. Proper hammer alignment is extremely important to prevent eccentric loadings on the pile. Otherwise driving energy transferred to the pile may be reduced considerably and structural pile damage due to excessive stresses near the top of the pile may result from eccentric loading. Leads can generally be classified as being either of the fixed or swinging type with several variations of each. Another less widely used type consists of a pipe or beam section that allows the hammer to ride up and down by means of guides attached to the hammer. When driving long slender piles, the use of intermediate pile supports in the leads may be necessary as long unbraced lengths may result in structural damage to the pile and may also contribute to violation of placement and driving tolerances. Leads are not absolutely necessary for every pile-driving operation, but they are normally used to maintain concentric alignment of the pile and hammer, and to obtain required accuracy of pile position and alignment while driving the pile, especially for battered piles. If leads are not required, a suitable template should be provided to maintain the pile in its proper location throughout the driving process. A brief description of fixed leads and swinging leads follows.

(a) Fixed Leads. Fixed leads, also called extended leads, are connected near the top with a horizontal hinge at the tip of the boom and extend somewhat above that point. Near the crane base, a spotter or horizontal brace is normally used and may be hydraulically operated to allow rapid achievement of pile batter. This combination provides maximum control, accuracy and speed when positioning the leads. A much more flexible version is the cardonic fully articulated lead, often called the swivel or three-way lead. Swivels are combined with moon beams or braces to allow movement not only in or out, but also side to side, and rotation of the leads. On large complex jobs which require the installation of a large number of battered piles, it is most advantageous to have leads capable of movement in all directions without having to reposition the entire driving rig. A special version of the fixed lead is the semi-fixed lead, in which the lead is free to move in the up and down direction independently of the crane boom. This type of lead is most beneficial when driving piles into a hole, ditch or over the edge of an excavation. An alternative to the semi-fixed lead is a fixed lead system accompanied by a pony or telescope lead, which secures the hammer in the fixed lead and allows driving below the bottom point of the fixed lead.

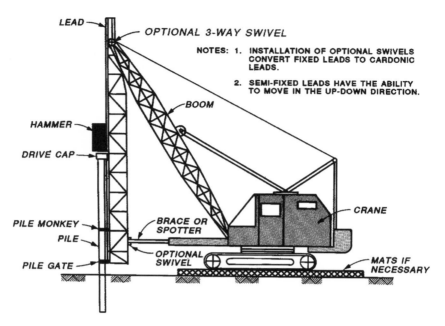

LEAD — | OPTIONAL 3-WAY SWIVEL

NOTES: 1. INSTALLATION OF OPTIONAL SWIVELS CONVERT FIXED LEADS TO CARDONIC LEADS.

2. SEMI-FIXED LEADS HAVE THE ABILITY TO MOVE IN THE UP-DOWN DIRECTION.

BOOM

HAMMER

DRIVE CAP

CRANE

PILE MONKEY

BRACE OR SPOTTER

PILE

OPTIONAL SWIVEL

PILE GATE

MATS IF NECESSARY

Figure 5-7. Typical fixed or extended leads

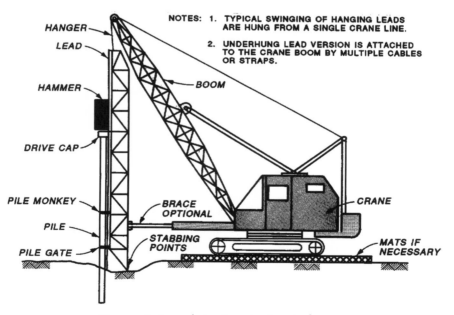

NOTES: 1. TYPICAL SWINGING OF HANGING LEADS ARE HUNG FROM A SINGLE CRANE LINE.

2. UNDERHUNG LEAD VERSION IS ATTACHED TO THE CRANE BOOM BY MULTIPLE CABLES OR STRAPS.

HANGER

LEAD

HAMMER

BOOM

DRIVE CAP

PILE MONKEY

BRACE OPTIONAL

CRANE

PILE

STABBING POINTS

PILE GATE

MATS IF NECESSARY

Figure 5-8. Typical swinging lead system

(b) Swinging Leads. The swinging lead, also known as the hanging lead, is hung from the crane boom by a single crane line and permits movement in all directions. A slightly different version is the underhung lead, which hangs from the boom itself by straps or pendant cables. Stabbing points are usually provided at the bottom end of the swinging lead for assistance when fixing position or batter. Swinging leads are lighter, simpler and less expensive than fixed leads, although precise positioning is slow and difficult. If swinging leads are to be used to drive piles that require a high degree of positioning accuracy, a suitable template should be provided to maintain the leads in a steady or fixed position. Leads that are not properly restrained may produce structural damage to piles, particularly concrete piles which are subject to spalling, cracking or even breakage. Swinging leads are especially useful to drive piles in a hole, ditch or over the edge of the excavation.

(4) Driving Caps. The drive cap will be defined here as a complete unit consisting of a helmet, anvil and cushion materials which function to properly transfer the driving energy from the hammer of the pile without damage to the pile. Various sites and types of helmets exist, two of which are shown in Figures 5-9 and 5-10. As impact hammers produce tremendous amounts of impact

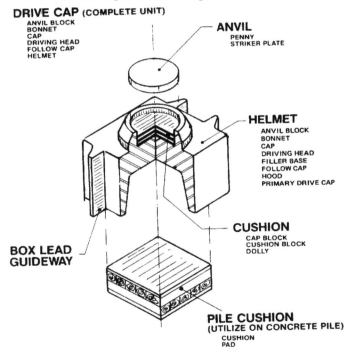

DRIVE CAP (COMPLETE UNIT)
ANVIL BLOCK
BONNET
CAP
DRIVING HEAD
FOLLOW CAP
HELMET

ANVIL
PENNY
STRIKER PLATE

HELMET
ANVIL BLOCK
BONNET
CAP
DRIVING HEAD
FILLER BASE
FOLLOW CAP
HOOD
PRIMARY DRIVE CAP

CUSHION
CAP BLOCK
CUSHION BLOCK
DOLLY

BOX LEAD GUIDEWAY

PILE CUSHION
(UTILIZE ON CONCRETE PILE)
CUSHION
PAD

Figure 5-9. Box lead mounting, air/steam and
diesel hammers

energy, the hammer blow must be transmitted uniformly over the top of the pile. Driving helmets made of cast steel are used for this purpose and are typically produced by the pile hammer manufacturer to suit its particular equipment. Experience indicates the helmet yields best results when guided by

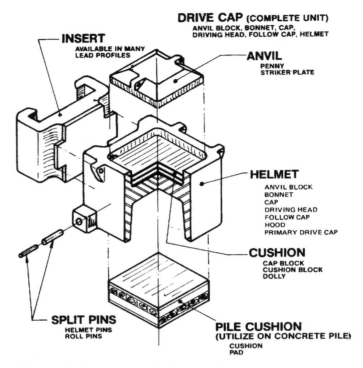

DRIVE CAP (COMPLETE UNIT)
ANVIL BLOCK, BONNET, CAP,
DRIVING HEAD, FOLLOW CAP, HELMET

INSERT
AVAILABLE IN MANY
LEAD PROFILES

ANVIL
PENNY
STRIKER PLATE

HELMET
ANVIL BLOCK
BONNET
CAP
DRIVING HEAD
FOLLOW CAP
HOOD
PRIMARY DRIVE CAP

CUSHION
CAP BLOCK
CUSHION BLOCK
DOLLY

SPLIT PINS
HELMET PINS
ROLL PINS

PILE CUSHION
(UTILIZE ON CONCRETE PILE)
CUSHION
PAD

Figure 5-10. Truss lead mounting, generally with
diesel hammers

the driving leads, although swinging helmets have proven satisfactory when
used with steel-H or heavy walled pipe piles. An appropriate helmet should
fit loosely around the pile top to prevent pile restraint by the helmet in
cases where the pile tends to rotate during driving. However, the fit should
not be so loose that it does not provide alignment of the hammer and pile.
While the helmet tends to protect the pile by distributing the blow, the
hammer may also require protection from the shock wave reflected back to the
hammer. For this purpose, a shock absorbing material known as the hammer
cushion is placed between the hammer ram and the helmet. Hammer cushions are
required for diesel hammers, while those powered by air or steam may or may
not require hammer cushions, depending on the particular hammer type and
manufacturer. The hammer cushion also serves to protect the helmet and the
pile. Commonly used hammer cushion materials are hardwoods, plywoods, woven
steel wire, laminated micarta and aluminum discs, and plastic laminated discs.
Thick blocks of hardwood are commonly used but have a tendency to crush, burn
and have variable elastic properties during driving. The laminated materials
are normally proprietary, provide superior energy transmission characteristics, maintain their elastic properties more uniformly during driving and have
a relatively long useful life. The use of materials such as small wood
blocks, wood chips, ropes and other materials which allow excessive loss of
energy or provide highly erratic properties should be discouraged (or pro-
hibited). Sheet asbestos has been commonly used in the past but is no longer
acceptable due to health hazards. A second cushion known as the pile cushion

is required when driving concrete piles. This cushion is placed between the helmet and the pile. The pile cushion protects the pile from compressive damage at the head of the pile and can also help control tensile stresses resulting from the tension shock waves produced by driving. Wood materials such as plywood and oak board are most commonly used. A pile cushion is rarely used when driving steel or timber piles. The type and thickness of the hammer and pile cushion materials have a major effect on the energy delivered to the pile. If the contractor chooses too soft a material, excessive energy absorption will result and driving may stall. On the other hand, choosing too hard a material will result in hammer or pile damage. Engineering experience combined with a wave equation analysis is the best method of selecting cushion materials and thicknesses. The complete drive cap design and properties of all components should be submitted by the contractor and reviewed for suitability. Cushion materials require periodic replacement during driving, since their effectiveness is reduced by excessive compression or deterioration. Indications of a need for replacement may be early throttling or bouncing of the hammer, or a ringing sound of the ram. The cushion design is based upon experience to a large extent, and the hammer manufacturer should be consulted in case of questions or distinct problems. Item 34 contains information regarding cushion properties and selection.

(5) Jetting Equipment. Typical equipment consists of jet pipes, a nozzle, pump, engine and hoses. The ensemble of equipment must be capable of providing the desired volume of water at the required pressure. Water volume and pressure must be sufficient to allow discharged water to surface along the sides of the pile. Typical pipe sizes range from 1.5 to 4.0 inches in diameter with nozzles approximately one-half the pipe diameter. Water pressures of 100 to 300 psi are most common but may run as high as 700 psi in isolated cases. Jetting pipes may be encased or cast into the pile, attached to the exterior of the pile or attached to the driving leads and thereby become movable. Moveable jets are preferable, if circumstances do not exclude their use, due to the relative high costs of permanently attached jets. The use of two jets, one on each side the pile, provides the most rapid penetration and best alignment control. When using multiple jets, each should be equipped with its own water source and both should be similarly operated at the same depths and pressures. A single jet placed on one side of the pile may result in excessive pile drift. Experienced personnel should be relied upon when selecting and sizing jetting equipment.

5-3. Pile Driving Studies. Pile driving studies are required for effective design of constructible pile foundations. When evaluating alternative pile types during the design phase, the designer must consider the effects of the pile installation method on the pile and soil capacities and on any existing structures in the proximity of the new foundation. The relative difficulty of driving the piles, and the procedure to determine when each pile has attained adequate capacity to end driving, must also be assessed. Past practices have addressed these considerations by use of empirical dynamic formulas, engineering experience and judgement, review of historical driving data, and various rules of thumb. More recently, the wave equation analysis and the dynamic pile driving analyzer methods have been generally accepted and should be employed. The pile-driving industry is presently moving toward exclusive use of wave equation analysis as the means for a designer to evaluate pile driveability, hammer selection, and limits of penetration. While the wave equation method provides superior analytical techniques, engineering experience and sound judgement are still very much a necessity. A review of pile installations for similar sites and structures can be extremely valuable in that regard. Rules of thumb can still be used for preliminary design and simple projects and should continue to be used during a design office's transition to the wave equation method. The designer must transform the results of these analyses into contract specifications that provide framework for the contractor to select appropriate equipment and installation procedures. Specifications should clearly define the basis of hammer approval, state criteria which will be used to establish the limits of penetration, and exclude installation methods or equipment deemed unsuitable. Analytical predictions are verified in the field by driving and static load tests, or the dynamic analyzer. Three principal topics are discussed in the following paragraphs; wave equation analysis, hammer selection, and penetration limitations. Wave equation results and penetration limitations can and should be used by field personnel to monitor and control the driving operation. In general, these topics are all interrelated.

a. Wave Equation. A wave equation analysis can provide the engineer with two very important items: first, a guide in the selection of properly sized driving equipment and piling to ensure the pile can be driven to final grade without exceeding allowable driving stresses; and secondly, a penetration rate expressed as a minimum number of blows per inch of penetration for impact hammers to determine when the pile has been driven sufficiently to develop the required capacity. This can be presented graphically by depicting the relationships between blows/inch (driving resistance) and ultimate static soil resistance (pile capacity) and blows/inch versus structural stresses in the pile. The graphs can then be used by field personnel and the contractor to monitor driving. When using the analysis results during installation, the design engineer must make certain that assumed design parameter values closely correspond to the actual values encountered in the field. This correlation can be accomplished by utilizing the load capacity and load transfer distribution obtained from static load tests and the dynamic analyzer. Analysis is based on a specific type and length of pile, and a driving system operating at an assumed efficiency in a modeled soil stratification. The results are applicable only to the assumed system and should only be used for the length of pile investigated. Incremental analysis is typically performed where the length of pile embedded into the ground is varied. Design application requires sound engineering judgement and experience where parameter (hammer, drive cap, pile and soil resistance) sensitivity is concerned. Research has shown that published hammer efficiencies (by the manufacturer) tend to significantly overestimate the energy actually absorbed by the pile in the field. Efficiency is also affected by placing the hammer on a batter and this can be a major source of error. Diesel hammers may have a variable stroke and a bracket analysis is strongly recommended. Hammer efficiency can be field-verified by good inspection techniques and more accurately by use of a dynamic pile analyzer. Data obtained from the wave equation analysis should be used with judgement for friction piles since pile set-up may occur. Data generated using the dynamic analyzer during original driving will not reflect pile set-up and may under-predict a pile's capacity. To produce data that reflect the true capacity of the pile, the pile should be restruck after set-up has occurred, usually a minimum of 14 days after initial driving. A wave equation analysis is recommended for all but the simplest of projects for which the designers have experience and should be performed for predicting behavior during design and confirming pile performance during construction of a project. The wave equation computer program "WEAP" (Wave Equation Analysis of Pile Driving) is available to Corps of Engineers offices. Item 34 contains an excellent discussion of wave propagation theory and its application to pile foundations.

  b. Hammer Selection.

  (1) General. Hammer selection may be the most important aspect of pile installation. In some installations only one hammer type may be applicable for the pile-soil combination, while for others several types may suitable. Evaluation must consider the need to use pile penetration rate as the means to end driving, the ability to drive the pile without structural damage or reducing soil capacity, the ability to obtain penetration rates within the desired band, and the realization that some hammer types may cause reduced capacities for identical pile lengths. In general, wave equation analysis supplemented by engineering experience and judgement should be the basis for hammer approval and criteria such as allowable driving stresses, desired penetration rates, and any other data used as a basis for approval that are clearly defined in the specifications. Wave equation analysis should normally be performed by the Government, and data that the contractor are required to submit must be clearly defined. Contractor disagreements with the Government's analysis can be contested by the contractor and resolved at his expense through resubmittals performed and sealed by a registered engineer, by field verification of driving and load tests, and by other methods approved by the design engineer.

  (2) Size selection for a particular hammer must consider the pile's anticipated driving resistance, ultimate capacity, pile stresses expected during driving, and pile set-up. The hammer type and size used for production should always match that used in the test program because a different hammer would most likely result in a different capacity. The designer or contractor may designate a number of hammers for the test program when warranted. Any changes in hammer type or size will usually require additional testing.

  (3) Prior to the wave equation method and development of the desk top computer, hammers were typically chosen based on dynamic formulas, rules of thumb, minimum energy rating based on pile type or load capacity, and methods which equated the pile weight to the weight of the moving hammer parts. These methods were primarily derived from experience and still have a place in hammer selection. However, these methods are only recommended as secondary procedures. Dynamic formulas are not recommended due to the lack of reliability and are considered to be inferior to the wave equation method. Table 5-1 is presented for information purposes only and to illustrate one of the many empirical methods still in use today. Tables such as this are generally being phased out and replaced by the wave equation method and sometimes supplemented by dynamic analysis in the field. These methods can and should still be utilized in an office in transition to the wave equation method.

  (4) Vibratory hammers require special attention as they have been shown to yield reduced capacity at work loads in some cases (Item 10, Item 15). Another reason for special attention is that there is no reliable way to evaluate driving resistance and driving induced stresses in piles as can be done for impact driven piles via pile driving analyzer and wave equation analysis. However, the potential economic advantage of a vibratory hammer cannot be discounted without adequate consideration, especially for large projects. Specifications can be written to require dual driving and load test programs if needed to address additional pile length and penetration

TABLE 5-1

SUGGESTED MINIMUM HAMMER ENERGY - IMPACT HAMMERS
(Taken from ARMY TM 5-849-1, May 1982)

Class I - Timber Piles

Capacity to 20 Tons   - 7,500 ft-lb

Capacity over 20 Tons to 25 Tons   - 9,000 ft-lb (Single-acting hammers)
                                       - 14,000 ft-lb (Double-acting hammers)

Capacity over 25 Tons   - 12,000 ft-lb (Single-acting hammers)
                                       - 14,000 ft-lb (Double-acting hammers)

Class II - Concrete and Steel Piles

Capacities to 60 Tons   - 15,000 ft-lb

Capacities over 60 Tons   - 19,000 ft-lb

---

limitations. Other engineering and construction agencies have permitted the use of a vibratory hammer but require a percentage of production piles be driven or struck with an impact hammer to determine relative capacity. In cases where tests indicate that additional pile length can be attributed to the hammer type, increased cost should be the responsibility of the contractor. The contractor may determine if the additional cost for testing and monitoring would be offset by increased production rate.

    c. Penetration Limitations. For impact hammers the rate of penetration is customarily defined as the blow count per unit length of pile penetration. Blow counts are typically recorded in the field on a per-foot basis until the pile approaches a designated tip elevation or the end of driving. At that point the blow count is usually recorded for each inch of penetration. Limiting penetration rates are designated to prevent overdriving, which may cause structural damage to the pile, and to provide guidance for determining the relative capacity attained during driving. Pile tip damage due to very difficult driving (commonly referred to as refusal) is not readily detectable when the pile encounters an obstruction or a hard bearing stratum prior to reaching the indicated tip elevation. Therefore, the limiting penetration rates, or the criteria necessary to determine limiting rates, should be specified. Rules of thumb, used to avoid structural damage, derived through experience and generally accepted by most engineers are listed in Table 5-2.

Table 5-2

Limiting Penetration Rates

Pile Type	Maximum Blow Count (blows per inch)
Timber	3-4
Concrete	10
Steel Pipe	10-20
Steel -H	10-20

The limiting penetration rates generally should be established by the Government and based upon results of wave equation analysis that have been correlated with results obtained from use of a pile driving analyzer during driving the test piles and the results of static load tests. Piles that derive their primary support from friction are driven to a predetermined tip elevation. For friction piles, the required length of penetration or tip elevation is determined from geotechnical data and capacity from test piles. The results of static load tests are then used to adjust the specified tip elevation or penetration length. Applicable penetration rate limits are compared with the rates encountered when driving the piles for the static load tests and adjusted if necessary. Piles that derive their primary support from end bearing in a hard soil layer or rock typically require a verification of load capacity, which may be indicated by the penetration rate. In this case the pile is normally driven to a specified blow count rather than a predetermined length. Once again, this blow count can best be obtained from wave equation analyses that have been correlated with driving and static load test data. Refinements in the wave equation analyses should be made by use of the pile dynamic analyzer when pile load test are not economically feasible. In either event the pile driving analyzer can be used to monitor the installation of piling. The designer should be wary that penetration rates observed in the field can easily be distorted by erratic or malfunctioning equipment and improper contractor operations. Distorted rates can be frequently attributed to an erratic or poorly maintained hammer, poor alignment of the hammer and pile, erratically behaving cushion materials, changing of a cushion near the end of driving, and noncontinuous driving that may allow the pile to set up and gain strength. A driven pile that has failed to acquire a specified tip elevation or penetration rate must be reanalyzed by the designer. If the safety factor for the pile or group is jeopardized, remedial measures are necessary, including extension of the driven pile by a splice, replacement of the pile with a longer one, or the addition of a sister pile. An end-bearing pile that stops short of its bearing stratum may be a candidate for special driving assistance, as discussed in paragraph 5-2a(3).

5-4. Control of Pile Driving Operations. Field installation requires continuous monitoring to ensure that an adequate foundation is achieved. All facets of installation require examination, from storage and handling to end of driving. If it is assumed that equipment is properly utilized and working at an efficient level, there remain two areas of concern: (1) monitoring installation to prevent structural damage, and (2) acquiring data to ensure that adequate capacity is obtained. Paragraph 5-3c previously discussed the use of wave equation analysis and selection of penetration limits in that regard. Field monitoring can be supplemented by dynamic analysis which can refine several assumptions made in the wave equation analysis (e.g. energy transfer to the pile), evaluate equipment performance, determine pile stresses estimated, and detect pile breakage. Piles suspected of sustaining structural damage or lacking in capacity can be further investigated by extraction or load testing.

    a.   Pile Driving Analyzers. These devices give a general indication of capacity, measure hammer and cushion performance and pile stresses from measurements of applied force and acceleration at the head of the pile. Capacity can often be inferred from the measurements using a simple damping constant for the soil. The soil damping constants can be calibrated from static load tests or by using special wave equation programs designed to infer capacity from pile-head measurements. The equipment is highly portable, performs most calculations on the job site, and requires trained and experienced personnel to operate. Analyzers are helpful to establish driving criteria and provide construction quality control when used in combination with static pile load test. The pile driving analyzer can be used in conjunction with theoretical predictions where static pile tests are not economically justified. Experience and sound engineering judgement are required when determining whether or not to use dynamic analyzers on a job, since this is a site-dependent decision. As previously stated, the analyzer only yields results of estimated capacity for the specific blow recorded, i.e., if data are taken during initial driving, the results can be distorted due to locked-in residual stresses, and any gain in capacity with time (set-up) is not accounted for. To account for the time-dependent gain in capacity, the pile should be restruck after a specified time (e.g. 7 to 14 days) has elapsed. If correlated with static pile tests and good driving records, the pile driving analyzer may be used successfully to predict capacity of production piles. It may also be used to indicate hammer efficiency, driving energy delivered to the pile or indicate pile breakage during driving. Specifications must address contractor and Government responsibilities when using a dynamic analyzer.

    b.   Records. Examples of the minimum records to be kept during driving are contained in Figures 5-11 and 5-12. The blow count per foot of pile penetration and the amount of free run drop under the hammer weight are two very obvious pieces of data to collect. When driving data are being analyzed, common questions are: the hammer type, manufacturer and any identifying numbers, has the hammer been modified in any way, was the hammer working at its rated capacity, cushion material and thickness, pile length and size, date of casting if precast concrete, depth of penetration, was driving continuous, were any special efforts of installation such as jetting or preboring applied, type of connection to the pile, magnitude of bias load, and the method and location of any splices. For vibratory hammers, the operating frequency, horsepower applied, and rate of penetration should also be recorded. Any occurrence of heave or subsidence for both the ground surface and adjacent piles should be noted. The method of hauling, storing, and handling the pile should also be recorded. Another item which should be recorded is whether or not the pile was properly handled as it was raised into the leads of the pile driver. Such records of data are invaluable when problems arise, performing as-built analysis and resolving contract disputes involving claims or litigation.

    c.   Proof Tests. Proof tests may become necessary if damage to a pile is suspected during handling or driving. Proof testing may also be prudent when large numbers of piles are driven into soils with highly variable stratification, and the driving records contain erratic data which can not be explained

PILE DRIVING REPORT

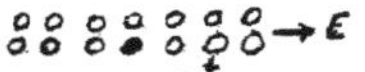

PROJECT _Golden Meadon Flood Gate_

CONTRACTOR _Bosick Construction_

HAMMER:

MAKE & MODEL _Conmaco 115_

WT. RAM _11500_   STROKE _3'_

ENERGY DELIVERED _37375_

DESCRIPTION AND DIMENSIONS OF
DRIVING CAP _14" x 14"_ (K) _1770 16_

SPEED: RATED   MEASURED _49_

STEAM OR AIR PRESSURE:
AT HAMMER _____ AT BOILER _128 psi_

JETTING PRESSURE AND ELEVATIONS:
Predrilling: Size & Depth

TIME: START DRIVING _14:49_   FINISH DRIVING _14:53_   DRIVING TIME _4 min_

INTERRUPTIONS (TIME, TIP ELEV. & REASON) _14:52_   _unhook piling_

PILE NO. _4_   _FS_

LOCATION _East Floodside_

TYPE: _Prestress Concrete_

DIMENSIONS _14 x 14 x 83.4_

LENGTH IN LEADS _100_

BATTER 1 ON (5)

ELEVATION OF GROUND _-14.0_

ELEVATION OF CUT-OFF _-4.98_

ELEVATION OF PILE TIP _-71.4_

ELEVATION OF SPLICES _None_

INSPECTOR _Fisher_   DATE _11/7/83_

DRIVING RESISTANCE

FT	NO. OF BLOWS	FT	NO. OF BLOWS	FT	NO. OF BLOWS	FT	NO. OF BLOWS	FT	NO. OF BLOWS	FT	NO. OF BLOWS	FT	NO. OF BLOWS	FT	BLOWS/INCH FOR LAST FOOT OF DRIVING — NO. OF BLOWS
0		15		30	2	45	5	60	2	75		90		1	4
1		16		31	3	46	5	61	2	76		91		2	6
2		17		32	3	47	5	62	2	77		92		3	6
3		18		33	3	48	5	63	2	78		93		4	8
4		19		34	3	49	4	64	2	79		94		5	8
5		20		35	4	50	3	65	3	80		95		6	8
6		21		36	5	51	4	66	2	81		96		7	10
7		22		37	7	52	3	67	3	82		97		8	10
8		23		38	6	53	3	68	3	83		98		9	10
9		24	1	39	10	54	3	69	3	84		99		10	10
10		25	1	40	10	55	3	70	2	85		100		11	10
11		26	1	41	11	56	2	71	3	86		101		12	10
12		27	2	42	12	57	2	72	3	87		102			
13		28	2	43	10	58	2	73	4	88		103			
14		29	2	44	5	59	2	74	100	89		104			

Figure 5-11.   Example of a completed Pile Driving Record

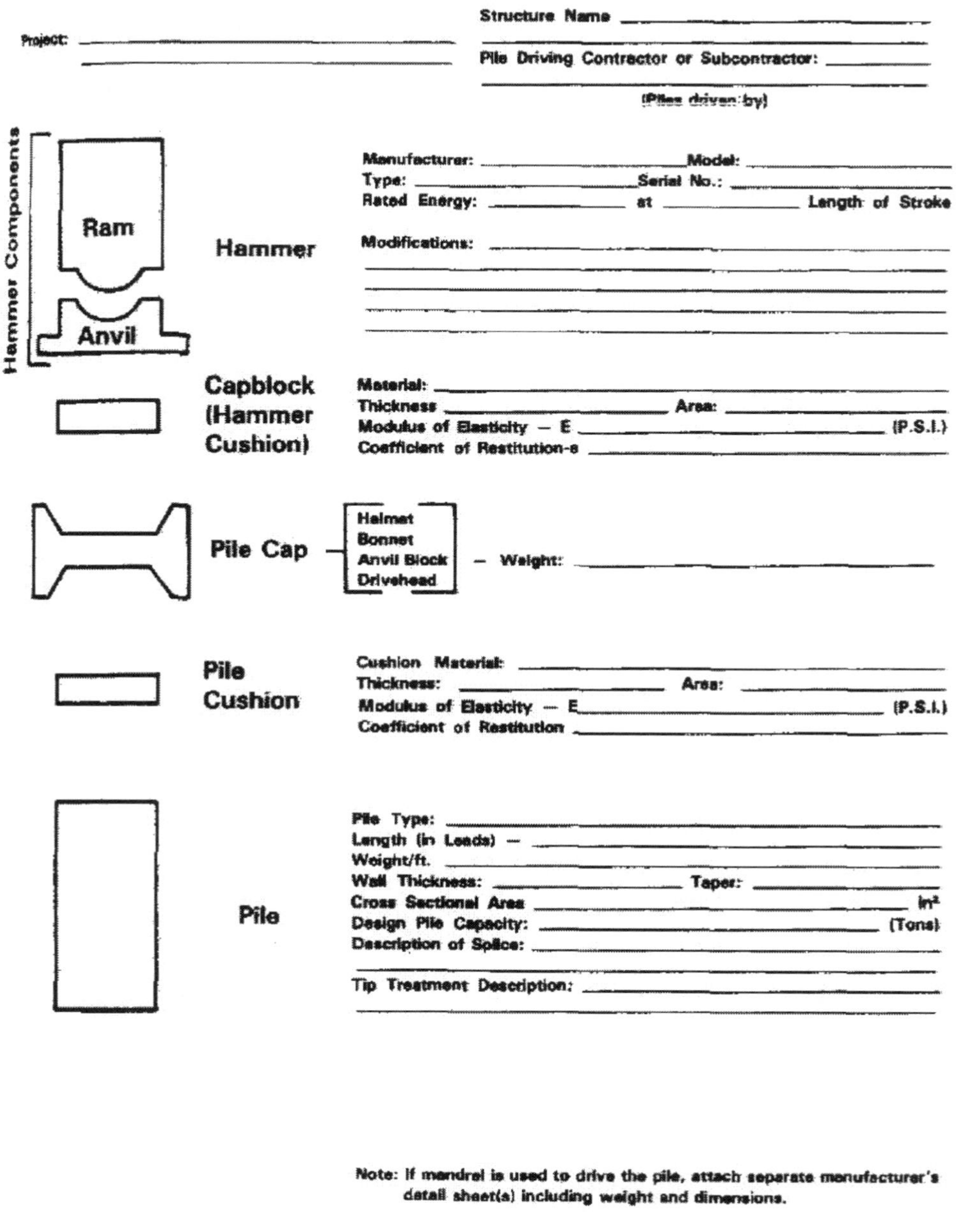

Project: _____

Structure Name _____

Pile Driving Contractor or Subcontractor: _____

(Piles driven by)

Hammer Components

Ram

Anvil

Hammer

Manufacturer: _____ Model: _____
Type: _____ Serial No.: _____
Rated Energy: _____ at _____ Length of Stroke

Modifications: _____
_____
_____
_____
_____

Capblock (Hammer Cushion)

Material: _____
Thickness _____ Area: _____
Modulus of Elasticity — E _____ (P.S.I.)
Coefficient of Restitution-e _____

Pile Cap

Helmet
Bonnet
Anvil Block
Drivehead

— Weight: _____

Pile Cushion

Cushion Material: _____
Thickness: _____ Area: _____
Modulus of Elasticity — E _____ (P.S.I.)
Coefficient of Restitution _____

Pile

Pile Type: _____
Length (in Leads) — _____
Weight/ft. _____
Wall Thickness: _____ Taper: _____
Cross Sectional Area _____ in²
Design Pile Capacity: _____ (Tons)
Description of Splice: _____

Tip Treatment Description: _____

Note: If mandrel is used to drive the pile, attach separate manufacturer's detail sheet(s) including weight and dimensions.

Figure 5-12. Example of a typical Pile Driving Equipment Report (Permission to reprint granted by Deep Foundations Institute (Item 31))

by the contractor's operations. An equally important indicator may be failure of a pile to reach the prescribed tip elevation or rate of penetration. Different types of proof testing can be employed depending upon the problem suspected. Testing with the pile driving analyzer may be performed by restriking previously driven piles, or data may be generated during driving and used in a wave equation analysis. A static quick load test, ASTM D1143-81 (Item 25), may also be used to determine the ultimate load carrying capacity of piles. On projects where it is anticipated that proof testing will be required, it is recommended that a line item be included in the bid schedule for performing such.

d. Extraction for Inspection. Piles are subject to structural damage during the driving process. Suspected damage below the ground surface would be cause for extracting a pile. Typical indicators are a pile suddenly drifting off location, erratic driving unexplained by the soil stratification, a sudden decrease in the driving resistance indicating breakage of the pile, or possible pile interference indicated by sound or vibration of nearby piles. Damage at the pile head may or may not indicate damage near the pile tip. If pile damage is suspected, the pile should be extracted and visually inspected. However, both the designer and field engineer should be cognizant of the fact that high costs and additional problems may be incurred as a result of extraction. For instance, a perfectly good pile may be damaged during the extraction procedure, particularly when extracting concrete piles, and soil stress states can be adversely modified around nearby piles where the subject pile is in a group. Costs associated with additional driving rig moves, obtaining and setting up extraction equipment, redriving time delays, and engineering and administrative costs are normally claimed by the contractor.

5-5. Results of Corps Experience.

a. Generalized Principles. The Corps of Engineers has been responsible for the design of construction of numerous foundations during the past 40 to 50 years. Through the efforts of Corps engineers, design and construction consultants, researchers, construction and individual contractors, the Corps has acquired vast experience in the foundation field. Advantage should be taken of this experience by researching available technical literature such as WES reports, engineering manuals, technical letters, project completion reports, contract specifications, design documents and verbal communications with other offices. Both the foundation design and constructability can benefit from this experience and historical data. Case histories of similar projects and similar sites are extremely valuable in this regard.

b. Case History. A typical case history for a recent project is presented in Appendix C.

c. Augmenting the Q/C and Q/A Processes. Providing for suitable controls during the construction process is an essential part of foundation design and contract preparation. Engineering judgement and past experience are required to determine the appropriate construction control procedures and methods for a particular type of pile foundation and soil system. The details of the construction controls should be developed during the foundation design process, tested during driving of test piles, and finalized upon evaluation of pile load test results. Proposed methods should be included in the design memoranda with proposed instrumentation and should reflect the functional importance and economic parameters of the project. An attempt should be made to anticipate and address all possible situations that may be encountered during pile driving and that could have a detrimental effect on the serviceability of the foundation. Adverse conditions are explained in paragraph 2-7 (and throughout this manual) and include pile material and geometry, subsurface conditions, driving equipment, and miscellaneous items.

(1) Pile material and geometry include defective pile material, strength, dimensions, straightness, spacing, alinement, location, etc.

(2) Subsurface conditions refer to strata variation, voids, liquefaction, obstruction, heave, densification, downdrag, water table, etc.

(3) Driving equipment pertains to defects in hammer, loads, accessories, etc.

(4) Miscellaneous items are vibration, adjacent structures or utilities, erratic values from PDA compared with wave equation results, etc.

5-6. As-Built Analysis.

a. Structural. Several variables may cause the actual pile foundation to differ from the initial design both in geometry (affecting pile loads) and pile capacities. The effects of these variations should be evaluated as discussed in paragraph 4-7i.

(1) Pile Geometry. As reflected by the relatively lenient driving tolerances normally allowed for pile position, orientation, and batter, physical control of the individual piles during driving is very difficult due to the nature of the large equipment required. Initial positioning and orientation of the pile in preparation for driving is not precise, and individual piles may have various amounts of initial camber and warp. During driving, variations in soil resistance combined with necessary clearances between the pile and the guides in the leads, and between the pile and hammer components (driving head or helmet, cushioning material, and hammer ram), permit variations in pile position and orientation. Small variations may be substantially amplified by long piles that are relatively flexible, by large pile batters, and by unexpected obstructions encountered in the soil during driving. Any combination of the aforementioned variables may result in differences between the design and the actual geometry of the resulting pile foundation. Variations in initial pile positioning and drifting of the pile during driving will each affect the final position and orientation of the head, the longitudinal axis, and the tip of the driven pile. The final position, orientation and batter of the pile head can be accurately measured after driving. However, the variation in orientation of the pile axis (curvature, twist, batter angle, and direction) and the pile tip elevation cannot be accurately determined after driving. If a pile is extracted for inspection and is undamaged, the axis orientation and tip elevation may be closely approximated for that pile as previously driven by extrapolation from measurements at the pile head.

(2) Pile Capacity. The pile capacity (axial, lateral, and buckling) is an interactive function of the properties of the soil and the pile, both governed by pile length. Also, the design length is determined by the batter angle and tip elevation. The batter angle may be affected by the unknown drift, which also affects the tip elevation. Variations in driving resistance may cause a substantial variation from design tip elevation. Based on

static pile load test results, a tip elevation is specified to provide the estimated design capacity and safety factor for the service piles. In addition, a minimum driving resistance (minimum blow count rate) required to develop the pile capacity and a maximum driving resistance that may be tolerated without structural damage to the piles are usually specified for guidance during driving. When the pile has been driven to the required tip elevation and the minimum driving resistance has not been developed, the pile may be extended by splicing and driven until the indicated driving resistance is developed, if deemed necessary. If the maximum driving resistance is developed prior to the pile's being driven to the design tip elevation, the situation must be investigated to determine the cause of the resistance (subsurface obstruction, gravel or cobbles, improper driving, etc.) When the cause has been determined, a decision must be made either to extract the pile and redrive it in another location, to leave the pile intact and cut off the upper portion, or to continue driving with a modified procedure or an increased maximum resistance parameter.

b. Geotechnical. As adjacent piles are generally driven into progressively denser materials, some piles driven previously may heave. Heave, which can be measured with quality surveying techniques, is detrimental to the performance of the pile foundation. Heave can be minimized by using the largest possible spacing between piles. Soil movements can be detected 5 feet to 8 pile diameters away from a pile, and a pile spacing of 3 diameters or less is not recommended. This problem can best be avoided by greater center-to-center spacing and driving radially outward from the center of the foundation. If significant heave occurs, the pile hammer should be replaced on the pile and the pile redriven. Since the pile can easily be damaged during this operation, the design engineer specifies driving energy or blow count that should not be exceeded. When the necessity for redriving develops during driving operations, the design engineer should evaluate and modify the driving sequence in an attempt to minimize the heave effect. When installation problems, especially heave, might conceivably occur while driving the piles, no pile head should be cut off until a sufficient number of piles have been driven or the driving operation has progressed for a sufficient distance to ascertain that problems will not be encountered or that the driving operation will no longer affect the driven piles. For any pile driven short of the specified tip elevation, the capacity should be recomputed and a safety factor estimated for the design load. If a significant number of piles, or a group of piles clustered together, are found to have less than the required safety factor, the structure should be reanalyzed using the recomputed capacity.

c. Wave Equation and Pile Driving Analyzer. Both of these are tools available to the pile foundation designer to evaluate his theoretical design from a constructability standpoint or to evaluate the as-built pile foundation. The pile driving analyzer is extremely useful in evaluating the field installation procedures. If used in conjunction with static load tests for correlation, it may be useful in evaluating the further installation of production piles. The pile driving analyzer, may be used to evaluate the pile hammer efficiency and to evaluate or detect potentially damaged piles. In using the pile driving analyzer, it should be noted that the analyzer uses dynamic theory to infer static pile capacity. In some soils the pile develops a significant portion of its ability to carry load after it has set-up for a period of time, therefore in such a case the pile should be restruck after this set-up has been allowed to occur. In general, a set-up period of 14 days is considered sufficient. The wave equation allows the dynamic analysis of the pile, soil, and driving equipment to be evaluated as a system, thereby allowing the designer to evaluate variables such as the pile cushion, the hammer, or even the pile material. The wave equation is a valuable tool that can be used to evaluate proposed methods for pile installation during design or construction, i.e., if the contractor proposes a hammer system to install a pile, this program can evaluate the data and aid in detecting potential problems or deficiencies.

5-7. Field Evaluation.

a. Driving Operations. The design engineers (structural and geotechnical) should visit the construction site while service piles are being driven to observe the driving operation and to investigate any difficulty that is encountered. The driving equipment should be inspected, and field conditions should be checked. Proposed methods for storage and handling of piles, positioning and aligning piles, supporting piles in leads, and transferring hammer energy to piles should be checked prior to driving the first pile. Driving of the first few piles should be observed to assure compliance with approved methods, proper operation of equipment (per manufacturer's rating) and proper driving procedures. In addition, the observer should watch for abnormal driving resistances and the occurrence of pile heave or voids adjacent to driven piles. When unusual difficulties develop, driving should again be observed and compared with the initial set of observations. The blow count for the piles should be plotted during the installation process to detect broken or damaged piles. Drastic drops in blow counts of similar piles in similar soils would be an indication of broken piles. It is also recommended that the blows per minute of hammer operation be recorded to indicate the efficiency of a hammer, since reduced rates often indicate reduced efficiency.

b. Pile Positioning. When pile driving has been satisfactorily completed, the actual position, orientation, and batter of each pile should be measured (extrapolating from head measurements) and compared with the design geometry. If substantial variations are found, an as-built analysis may be required, as discussed in paragraph 4-7i. If there is substantial variation from the design tip elevation or from the anticipated driving resistance, the pile capacities should be re-evaluated. The piles may be inspected for drifting, which would be evidenced by voids adjacent to the pile. Drifting could be caused by striking a hard underground object or another pile. (A change in the impact sound of the pile during driving can be used to detect piles striking an obstruction.) It is also recommended that the blows per minute of hammer operation be recorded to indicate the efficiency of a hammer. The blow count records should be studied while driving is ongoing.

c. Static Loading. Prior to static load test, the jack and load cell should be checked, and the load settlement data should be plotted and checked in the field during the test.

CHAPTER 6

FIELD PILE TESTS

6-1. _General_.

a. A field pile test program will generally consist of two types of procedures: load tests to determine the load capacity of service piles; and
driving selected types of piles with selected types of hammers and recording
data on driveability. These tests may be conducted separately or, as is more
common, concurrently. The latter can be accomplished by simply recording the
necessary data during the driving of the piles that will be load tested.
Field pile tests are performed to verify or predict driving conditions and/or
load capacity of service piles at the construction site. Verification is the
process of test driving and loading the designated piles to predetermined
static loads to confirm prior design capacity calculations that were based on
static or dynamic type equations, previous experience, or empirical methods.
This process is used primarily only to confirm the load capacity and driveability of the selected pile as service piles. Prediction is the process of
test driving various piles or of loading test piles in increments to failure
to determine the length, type, and ultimate capacity of the service piles at
the construction site. Prediction tests differ from the verification process
in that these tests will be utilized for design purposes. Therefore, the
final pile size, type, and lengths generally have not yet been determined when
these tests are conducted. Prediction type tests are more common to large or
major structures where changes in pile lengths, size, type, or method of
installation could result in significant economic savings due to the large
number of piles involved.

b. A field load test on test piles may consist of two types, axial
and/or lateral. These tests may be performed on single piles or pile groups.

6-2. _Decision Process_. Pile tests are not practical or economically feasible
under certain circumstances, but they are always technically desirable. In
the initial design stages, as soon as the requirement for a pile foundation is
confirmed, several factors should be considered and evaluated to govern the
decision process.

a. Factors for Consideration. Appropriate evaluations of the following
factors should be included in the reconnaissance study and updated in each
successive report stage for the project.

(1) Engineering. The size, type, and importance of the proposed structure, type of subsurface conditions, and economics are engineering factors.
Size and economics are directly related; since pile tests are relatively expensive, a structure requiring a small number of piles could not normally
justify the expense of a pile test. The type and functional importance of a
structure could offset the added cost of a pile test program for a complex
foundation when the consequences of a potential failure would be catastrophic,
especially if the information obtained from subsurface investigations indicated unusual conditions that would be difficult to interpret. The costs of
pile tests should be compared with the potential project savings from basing
the foundation design on the test results with reduced safety factors. Also,
when a requirement for a field pile test program has been established, the

technical aspects, scheduling, estimated costs and any intangible benefits of the two following alternatives should be evaluated:

(a) Perform pile tests by separate contract and complete the foundation design prior to award of the construction contract.

(b) Award a construction contract with a selected type or types of piles and estimate pile lengths with a pile test program included to determine the actual pile lengths required.

(2) Budget and Schedules. During the reconnaissance study phase of a new project, a preliminary foundation evaluation for each potential project site is required to determine the need for pile foundations. For each potential site that needs a pile foundation, initial cost estimates and schedules must identify and include resources for necessary site investigations to provide adequate data for proper site selection and estimated costs for the feasibility study phase. The feasibility report should include a recommended site, preliminary alternate foundation designs, and the scope of required pile tests, etc. Required resources, schedules and cost estimates must be revised for each successive report phase to reflect the current design status of each project component. If a separate contract for pile tests is recommended prior to award of the construction contract, appropriate adjustments must be made to the schedules and budgeted funds.

(3) Available Test Site. Consideration has to be given to the timing of the test pile program in relation to the construction schedule. The test pile program may be a separate contract awarded and completed before construction begins. The advantages of this approach are many: the pile size, type, lengths, and preferred installation method can be determined before construction of the project; any problems with the pile test and potential problems with design revealed by the pile tests may be resolved prior to construction when they are more likely to be less costly. The disadvantages of this approach are: the design schedule is extended to allow time for the separate operations; the test conditions may not closely simulate design assumptions since excavation, water conditions, fill, etc. may not necessarily match construction conditions; and additional problems may develop if different contractors and/or equipment are used during the testing program and driving the service piles. These advantages and disadvantages must be evaluated in relation to the site availability.

(4) Site access. If the decision is made to conduct the testing program during the construction process, the scheduling of the test program becomes important. The tests must be conducted early in the construction process, since the contractor generally must await the outcome of the tests before ordering the service piles. However, the testing cannot be scheduled too early, since the test site needs to be prepared and accessible.

b. Proof Test. In the overall design process, field tests are normally scheduled after some estimate of pile capacity and driveability has been made. The driveability of piles is generally evaluated early in the design process, usually in the General Design Memorandum stage when basic design decisions are being made relative to the foundation. Also at this point, an estimate of pile capacity is made. During the Feature Design Memorandum phase of design, a more accurate prediction of pile capacity and/or required pile lengths is made, and a test pile program is established to verify the design assumptions and pile driveability. The final stage of the design process is the actual testing program. As discussed earlier, the results of this testing program may be used solely to verify the predicted pile capacities, and/or required pile lengths, or may be used as an extension of the design process by changing the pile size, type, lengths, or installation method of the service piles as required. Refer to preceding paragraphs relative to the timing of field tests with respect to construction, site availability, site access, and potential problems.

6-3. Axial Load Test.

a. Compression.

(1) General. The load test should basically conform to the procedures contained in ASTM D1143 (Item 25). This standard is recommended as a guide and should be modified as required to satisfy individual project requirements. Important aspects of the test are discussed in the following paragraphs. Information on conducting dynamic load tests may be obtained from the Geotechnical Laboratory of the US Army Engineer Waterways Experiment Station in Vicksburg, MS.

(2) Applying Load. In the past, test loads were generally applied by placing dead weight as a vertical load directly on top of the piles to be tested. In current practice, loads are applied by jacking (with a hydraulic ram) against a stable, loaded platform, or against a test frame anchored to reaction piles. Typical loading arrangements are illustrated in the referenced ASTM D1143 Standard. If reaction piles are used, various studies have indicated that the distance between these piles and the test piles should be a minimum of at least 5 pile diameters for nondisplacement piles to up to 10 diameters for displacement piles.

(3) Jacks and Load Cells. Most load tests are conducted with hydraulic jacks to apply the load, and load cells to measure the load. The hydraulic jack ram or the load cell should have a spherical head to minimize eccentricity between the jack and the loading frame. Both the jack (with pressure gage) and the load cell should be calibrated by a qualified lab to include calibration curves. During the load test both the load cell and the jack pressure gage should be read and compared. In the event it is later discovered or determined that the load cell has malfunctioned, the pressure gage readings will then be available. It is also important during the test to continually monitor the load cell to ensure that the load increment is being maintained at a constant value. The load has a tendency to decrease due to pile penetration into the ground, deflection of the test beams, and loss of hydraulic fluid from leaking valves, etc.

(4) Measuring Devices. Settlement measurement devices are usually dial gages, with some other independent system as a backup. It is important to protect these devices from the weather, since direct sunlight can significantly affect the readings. The ASTM D1143 Standard illustrates a typical installation. Consideration should also be given to measuring the lateral movement of the pile during the test procedure to detect eccentric loading conditions. All dial gages should be calibrated by a qualified lab.

(5) Instrumentation. If the distribution of the pile load along its embedded length is required, and not merely the total or ultimate load, the pile

should be instrumented with telltales or strain gages. All instrumentation should be tested after the pile is driven to verify that it is still functioning properly. Refer to the ASTM D1143 Standard for possible telltale arrangements. A strain-gage system may also be used. These systems, although generally more expensive, but can yield excellent results if properly installed and read.

(6) Net Settlement. Provisions should be made during the pile tests to determine the net settlement of the pile (i.e. the total settlement less the elastic compression of the pile and soil). This is required to develop a net settlement (i.e. the pile tip movement) versus load curve to determine pile capacity. Net settlement may be determined by loading and unloading the pile in cycles (see the ASTM D1143-81 Standard (Item 25)) by employing a telltale located at the pile tip.

(7) Technical Specifications. Plans and specifications for the pile test should be developed generally in accordance with the referenced ASTM D1143 Standard and should be specifically modified as needed to satisfy the particular project requirements and subsurface conditions. Technical specifications should include the following as a minimum:

(a) Type, size, length, and location of pile(s) to be tested

(b) Size and capacity of pile driving equipment

(c) Driving criteria and any special installation methods required

(d) Types of tests to be conducted and maximum testing capacity necessary

(e) Required testing equipment and instrumentation, including calibration, to be furnished

(f) Testing procedures to be followed

(g) Data to be recorded and reported

(h) Report format to be followed

(i) Provisions for additional tests

(j) Payment schedule and schedule of bid items

b. Tension.

(1) General. The load test should basically conform to the procedures contained in ASTM D3689-78 (Item 23). This standard is recommended as a guide, and should be modified as required to satisfy individual project requirements. Important aspects of the test are discussed in the following paragraphs.

(2) Testing. Tension tests are often conducted on piles which have previously been tested in axial compression. Some advantages to this are: a direct comparison of tension and compression on the same subsurface profile, cost savings in not having to drive an additional pile, and information on

piles that must function in both tension and compression under operating
conditions. Some disadvantages are: residual stresses may significantly
affect the results, remolding of the soil may take place during the first
test, and a waiting period is generally required between the compression and
tension test. Appropriate references should be consulted relative to residual
stresses and the necessary waiting period (paragraph 6-3d(5)).

(3) Other Considerations. The same aspects of compression tests need to
be considered for conducting tension tests: applying the load, use of jacks,
load cells, measuring devices and instrumentation, net settlement, and technical specifications. Paragraph 6-3a and the referenced ASTM Standards must be
referred to for compression and tension tests. Interpretation of test results
is described in paragraph 6-3e.

    c. Quick Load Tests.

(1) General. The load test outlined in the referenced ASTM
D1143 Standard (Item 25) can be described as a slow, maintained-load test.
The duration of this procedure can exceed 70 hours or longer, especially when
cyclic loading is included. There are alternatives to this "slow" test when
time is of the essence, or when a quick method is needed to verify service
pile capacities that were projected from test piles using the slow,
maintained-load method. These alternatives are the constant-rate-of-
penetration test (CRP), and the quick, maintained-load test. Both of these
methods are permissible options described in the referenced ASTM Standard
(Item 25). The CRP test appears to be seldom used in this country. The
quick, maintained-load test is often used, although its format has many
variations.

(2) CRP Test. In a CRP test, the load is applied to cause the pile head
to settle at a predetermined constant rate, usually in the vicinity of
0.01 inch per minute to 0.1 inch per minute, depending on whether the subsurface conditions are cohesive or granular, respectively. The duration of
the test is usually 1 to 4 hours, depending on the variation used. The particular advantage of the CRP test is that it can be conducted in less than one
working day. A disadvantage is that ordinary pumps with pressure holding devices like those used for "slow" tests are difficult to use for the CRP test.
A more suitable pump is one that can provide a constant, nonpulsing flow of
oil. Appropriate references should be consulted relative to the CRP test, if
it is utilized.

(3) Quick Maintained-Load Test. In this test, the load is applied in
increments of about 10 percent of the proposed design load and maintained for
a constant time interval, usually about 2 to 15 minutes. The duration of this
test will generally be about 45 minutes to 2 hours, again depending on the
variation selected. The advantage of this test, like the CRP test, is that it
can be completed in less than 1 working day. Also, unlike the CRP test, no
special equipment is required. Appropriate references should be studied if
this test is utilized.

(4) Other Considerations. The same aspects of axial load testing discussed in paragraphs 6-3a and 6-3b, and the referenced ASTM Standards
(Items 23 and 25), need to be considered for "quick" tests. These include
applying the load, use of jacks, load cells, measuring devices and

instrumentation, net settlement, and technical specifications, when applicable. Refer to paragraph 6-3e for interpretation of test results.

    d. General Considerations.

    (1) Reaction Pile Effects. If a pile is loaded by jacking against reaction piles, the effects of the reaction piles on the test pile (increase or decrease in intergranular pressures) should be taken into consideration.

    (2) Load to Failure. Test piles should be loaded to failure when possible, as this test yields valuable information to the designer. Ideally, care must be taken as failure is approached to collect data more frequently than at sub-failure loads and to maintain the same rate of loading employed before reaching failure.

    (3) Pile Driving Analyzer or Quick Tests. On a large or significant structure, consideration should be given to using a pile driving analyzer and/or performing quick tests in conjunction with the regular test pile program. The additional information provided (hammer efficiency, etc.) could aid the designer in evaluating the adequacy of the service piles, should unexpected problems develop, without having to conduct additional and costly conventional pile testing. In conjunction with the pile driving analyzer, a wave equation analysis should be performed prior to the pile test to calibrate the wave equation analysis with the results of the pile driving analyzer. Refer to Chapter 5, paragraph 5-3a for a discussion of the wave equation analysis. The pile driving analyzer is discussed in paragraph 5-4a. On a smaller or less significant project, the use of a pile driving analyzer may not be economically justified. However, a wave equation analysis is very simple to run and should be performed.

    (4) Location of Test Site. Test piles should be located as near as possible to a boring. In many instances, circumstances warrant that a boring be taken specifically for the pile test. Piezometric data should also be available. Conditions measured by the piezometers should be correlated with design/operating conditions.

    (5) Waiting Period. The waiting period between the driving of the test piles and the pile load test should allow sufficient time for dissipation of excess pore water pressures resulting from the pile driving operation. If sufficient time is not allowed, the test results may mislead the engineer to select pile capacities that are lower than the actual values. This waiting period is a function of many interrelated complex factors and can significantly affect the results of the pile test. Generally, piles driven into cohesive foundations require more time than those placed in granular materials. For the cohesive case, 14 days is recommended. An absolute minimum of 7 days is required. The referenced ASTM Standards (Items 23 and 25) have some guidance in this area. Other references should be studied before writing the technical specifications, if applicable. For cast-in-place concrete piles, the waiting period should consider the curing time and resulting strength of the concrete, and the possible effects of concrete hydration on the soil surrounding the pile.

    (6) Reporting Test Results. Data from the load tests should be recorded and reported in an orderly fashion. Items to be included are listed in the referenced ASTM Standards (choose only those that are applicable to the

project requirements). Results will often be used to estimate axial capacity and/or driving characteristics for other projects with similar subsurface conditions, or expansion/modification of the existing project. Thus, it is important to maintain records in a manner that will be useful and easy to interpret at some future date.

e. Interpretation of Results. The interpretation of the tests results generally involves two phases, analyzing the actual test data, and application of the final test results to the overall design of the service piles and the structure.

(1) Axial Capacity Determination Method. There are many empirical and arbitrary methods available to determine the axial capacity of a pile from load test data. Some of these methods are contained in bibliographical material found in Appendix A. It should be noted that the methods described in Table 6-1 are for informational purposes only, are not necessarily current practice, nor necessarily recommended by the respective listed sources. The methods are listed merely to indicate historical practice and the diversity of philosophy.

(a) Corps of Engineers Method. The following method has often been used by the Corps of Engineers and has merit: determine the load that causes a movement of 0.25 inch on the net settlement curve; determine the load that corresponds to the point at which the settlement curve has a significant change in slope (commonly called the tangent method); and determine the load that corresponds to the point on the curves that has a slope of 0.01 inch per ton. The average of the three loads determined in this manner would be considered the ultimate axial capacity of the pile. If one of these three procedures yields a value that differs significantly from the other two, judgment should be used before including or excluding this value from the average. A suitable factor of safety should be applied to the resulting axial pile capacity. See Figure 6-1 for an example of this method.

(b) Davisson Method. A commonly used method to evaluate pile tests is one suggested by Davisson (Item 30). The failure load is defined as the point at which the movement of the pile butt exceeds the elastic compression of the pile by 0.15 inch plus a factor (B/120) that is a function of pile diameter (B). The advantage of this method is the inclusion of physical pile size (length and diameter) in the definition of failure load. This failure load should exceed two times the allowable load.

(2) Foundation Considerations. Once the axial capacity of the pile is established, the next step is to interpret this information relative to the known foundation conditions, the nature of the loads on the structure, the size and significance of the structure, and any other pertinent information.

(a) Group Effects. It should be noted that the results of a single pile test may not indicate the capacity of a group of similar piles. The effects of group loading are experienced much deeper in the foundation than those of the single pile. Group loading may result in the consolidation of a soft clay layer that would otherwise be unaffected by a single pile loaded to the same unit load.

(b) Settlement. A pile test on a single pile will generally yield sufficient data to determine the failure load or bearing capacity. However, the

Table 6-1

Methods of Pile Load Test Interpretation

---

1. Limiting Total Butt Settlement

    a. 1.0 in. (Holland)
    b. 10% of tip diameter (United Kingdom)
    c. Elastic settlement + D/30 (Canada)

2. Limiting Plastic Settlement

    a. 0.25 in. (AASHO, N.Y. State, Louisiana)
    b. 0.5 in. (Boston) [complete relaxation of pile assumed]

3. Limiting Ratio: Plastic/Elastic Settlement

    1.5 (Christiani and Nielson of Denmark)

4. Limiting Ratio: Settlement/Unit Load

    a. Total   0.01 in./ton (California, Chicago)
    b. Incremental 0.03 in./ton (Ohio)
                    0.05 in./ton (Raymond International)

5. Limiting Ratio: Plastic Settlement/Unit Load

    a. Total   0.01 in./ton (N.Y. City)
    b. Incremental 0.003 in./ton (Raymond International)

6. Load-Settlement Curve Interpretation

    a. Maximum curvature - plot log total settlement vs log load; choose
                        point of maximum curvature
    b. Tangents - plot tangents to general slopes of upper and lower portion
                    of curves; observe point of intersection
    c. Break point - observe point at which plastic settlement curve breaks
                        sharply; observe point at which gross settlement curve
                        breaks sharply (Los Angeles)

7. Plunge

    Find loading at which the pile "plunges," (i.e., the load increment could
    not be maintained after pile penetration was greater than 0.2 B).

8. Texas Quick Load

    Construct tangent to initial slope of the load vs gross settlement curve;
    construct tangent to lower portion of the load vs gross settlement curve
    at 0.05 in./ton slope; the intersection of the two tangent lines is the
    "ultimate bearing capacity."

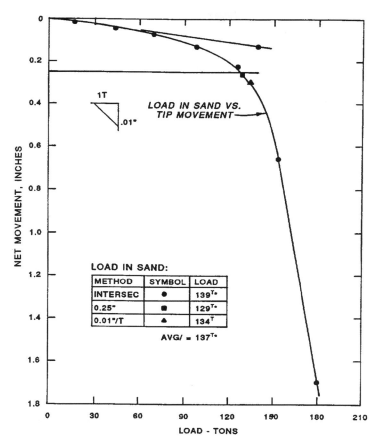

Figure 6-1. Corps of Engineers method for
determining axial capacity

pile test does not provide data relative to the settlement of the pile under
operating conditions in a cohesive foundation. The load test is generally
conducted in too short a time frame to enable clays to consolidate. Therefore, a significant amount of settlement may occur during the life of the
structure--settlement that would not be predicted by the pile test. For granular foundations however, the pile test does generally yield adequate data on
bearing capacity and settlement.

(c) Operating Conditions. Consideration should be given to the possibility that the test pile conditions may differ significantly from the operating conditions for the structure. Examples are: potential uplift in pervious
strata that are dewatered during the pile test; backfill or excavation after
the pile test; the nature of the loading on the piles (static, dynamic, long
term, short term, etc.); battered service piles in lieu of vertical test
piles; lateral loading effects; and negative skin friction.

(d) Driving Effects. The effects of driving many service piles may change the conditions existing during the test. Piles driven into a granular material may densify the foundation and increase pile capacity, while piles driven into a sensitive cohesive foundation may decrease pile capacity.

(e) Layered Foundations. Layered foundations may cause service piles to perform differently than indicated by test piles. During a test pile loading, a cohesive layer may be supporting the piles. During the life of the service pile, this same cohesive layer may consolidate under the load, and transfer the load to another soil layer not stressed during the pile test. The service pile may also shift from being a friction pile to being more of an end bearing pile under similar conditions.

(f) Residual Stresses. Residual stresses that may be present during the pile test may be significant and must be considered. These stresses may be detected by instrumenting the piles and taking readings prior to and just after driving. If residual stresses are present, it may be necessary to consider these stresses when evaluating the distribution of the tip and skin resistance.

(g) Tip Elevations. Finally, if the pile tests are used to project pile capacity for tip elevations other than those tested, caution should be exercised. In a complex or layered foundation, selecting a tip elevation for the service piles different from the test piles may possibly change the pile capacity to values other than those projected by the test. As an example, shortening the service piles may place the tips above a firm bearing stratum into a soft clay layer. In addition to a loss in bearing capacity, this clay layer may consolidate over time and cause a transfer of the pile load to another stratum. Lengthening the service piles may cause similar problems and actually reduce the load capacity of the service piles if the tips are placed below a firm bearing stratum. Also, extending tips deeper into a firmer bearing stratum may cause driving problems requiring the use of jetting, predrilling, etc. These techniques could significantly alter the load capacity of the service piles relative to the values revealed by the test pile program and should be considered in setting tip elevations for service piles.

6-4. Monotonic Lateral Load Test.

a. General. The main purpose of a lateral load test is to verify the values of n_h or E_s used in design. The value of the cyclic reduction factor used in design can also be verified if the test pile is cyclicly loaded for approximately 100 cycles. The basis for conducting a lateral load test should be ASTM D3966-81 (Item 24) modified to satisfy the specific project requirements.

b. Applying Load. A lateral load test is most easily conducted by jacking one pile against another. In this manner, two lateral load tests can be conducted simultaneously. When applying the lateral load to the pile, it is important to apply the load at the ground surface with no restraint at the pile head. This gives a free-head pile boundary condition and makes the data easy to reduce to curves of n_h or E_s versus pile top deflection. The loads are applied with a hydraulic jack. A spherical bearing head should be used to minimize eccentric loading.

c. Instrumentation. The minimum amount of instrumentation needed would be dial gages to measure lateral pile head deflection and a load cell to measure applied load. A load cell should be used to measure load instead of the pressure gage on the jack because pressure gage measurements are known to be inaccurate. Additional instrumentation could consist of another level of dial gages so the slope at the top of the pile can be calculated, and an inclinometer for the full length of the pile so that lateral pile deflection at any depth along the pile can be calculated. If p-y curves are necessary for the pile foundation design, then strain gages along the pile to measure bending moment are needed. However, since the purpose of lateral load tests described in this section is to verify or determine pile-soil properties to be used in the normal design of a civil works project, the use of strain gages along the length of the pile is not recommended. Accurate strain-gage data are difficult to obtain and only of value in research work where p-y curves are being developed. Strain gages should not be installed by construction contractors because they do not have the necessary expertise to install them. If strain gages are used, consultants experienced in their use should be hired to install them, and record and reduce the data.

d. General Considerations.

(1) Groundwater. The location of the ground-water table has an effect on how laterally loaded piles behave. For this reason it is important to have the groundwater table during testing as near as possible to the level that will exist during operation of the structure.

(2) Load to Failure. It is important to carry the load test to failure. Failure is defined as when the incremental loading can not be maintained.

(3) Location of Test Site. Piles should be located as near to the site of the structure as possible and in similar materials.

(4) Reporting Test Results. Accurate records should be made of the pile installation, of load testing, and of the load test data to document the test.

6-5. Cyclic Lateral Load Test.

a. General. The main purpose of a cyclic lateral load test is to verify the value of the cyclic loading reduction factor (R_c) used in design. Approximately 100 cycles of load should be used in a cyclic load test. The load test should be conducted according to ASTM D3966-81 (Item 24) modified as necessary for cyclic loading and specific project requirements. The instrumentation, equipment, and test layout necessary for conducting the cyclic load test is similar to that required for the monotonic lateral load test.

b. Procedure. Generally the cyclic lateral load test would be done in combination with the monotonic lateral load test on the same piles. Since repeated testing of the same pile can cause permanent nonrecoverable deformations in the soil, the sequence of testing is important. One sequence for doing the monotonic and cyclic lateral load test is outlined as follows: The designer must first select the load level of the cyclic test. This may be done from a known load level applied to the pile founded monoliths or a deflection criterion. A deflection criterion would consist of loading the load test piles to a predetermined deflection and then using that load level for the cyclic load test. Using the cyclic load level, the test piles would be cyclically loaded from zero loading to the load level of the cyclic load test. This cyclic loading procedure would be repeated for the number of cycles required. Dial gage readings of lateral deflection of the pile head should be made at a minimum at each zero load level and at each maximum cyclic load level. Additional dial gage readings can be made as necessary. After the last cycle of cyclic loading has been released the test piles should then be loaded laterally to failure. That portion of the final cycle of load to failure above the cyclic test load can be superimposed on the initial cycle of loading to get the lateral load-deflection curve of the piles to failure.

APPENDIX A

REFERENCES

A-1. <u>References are US Army Corps of Engineers Directives.</u>

(1)   TM 5-849-1, "Pile Driving Equipment."

(2)   EM 385-1-1, "Safety and Health Requirements Manuals."

(3)   EM 1110-2-1902, "Stability of Earth and Rock Fill Dams CH 1."

(4)   EM 1110-2-1906, "Laboratory Soils Testing CH 1-2."

(5)   EM 1110-2-1910, "Inspection of Earthwork Construction."

APPENDIX B

BIBLIOGRAPHICAL AND RELATED MATERIAL

B-1. <u>Corps of Engineers Publications.</u> All related publications are available on interlibrary loan from the Research Library, US Army Engineer Waterways Experiment Station, Attn: CEWES-IM-MI-R, 3909 Halls Ferry Road, Vicksburg, MS 39180-6199.

1. Brown, D. A., and Reese, L. C. 1988 (Feb). "Behavior of a Large-Scale Pile Group Subjected to Cycle Lateral Loading," Miscellaneous Paper GL-88-2, US Army Engineer Waterways Experiment Station, Vicksburg, MS.

2. CASE Task Group on Pile Foundations. 1980 (Dec). "Basic Pile Group Behavior," Technical Report K-80-5, US Army Engineer Waterways Experiment Station, Vicksburg, MS.

3. CASE Task Group on Pile Foundations. 1983 (Sep). "Basic Pile Group Behavior," Technical Report K-83-1, US Army Engineer Waterways Experiment Station, Vicksburg, MS.

4. Dawkins, William P. 1984. "User's Guide: Computer Program for Soil-Structure Interaction Analysis of Axially Loaded Piles (CAXPILE)," Instruction Report K-84-4, US Army Engineer Waterways Experiment Station, Vicksburg, MS.

5. Hartman, Joseph P., Jaeger, John J., Jobst, John J., and Martin, Deborah K. 1989 (Jul). "User's Guide: Pile Group Analysis (CPGA) Computer Program," Technical Report ITL-89-3, US Army Engineer Waterways Experiment Station, Vicksburg, MS.

6. Jaeger, John J., Jobst, John J., and Martin, Deborah K. 1988 (Apr). "User's Guide: Pile Group Graphics (CPGG) Computer Program," Technical Report ITL-88-2, US Army Engineer Waterways Experiment Station, Vicksburg, MS.

7. Martin, D. K., Jones, H. W., and Radhakrishnan, N. 1980 (Jun). "Documentation for LMVDPILE Program," Technical Report K-80-3, US Army Engineer Waterways Experiment Station, Vicksburg, MS.

8. Morrison, C. S., and Reese, L. C. 1988 (Feb). "A Lateral-Load Test of Full-Scale Pile Group in Sand," Miscellaneous Paper GL-88-1, US Army Engineer Waterways Experiment Station, Vicksburg, MS.

9. Mosher, R. L. 1984 (Jan). "Load-Transfer Criteria for Numerical Analysis of Axially Loaded Piles in Sand, Part 1: Load-Transfer Criteria," Technical Report K-84-1, US Army Engineer Waterways Experiment Station, Vicksburg, MS.

10. Mosher, R. L. 1987. "Comparison of Axial Capacity of Vibratory-Driven Piles to Impact-Driven Piles," Technical Report ITL-87-7, US Army Engineer Waterways Experiment Station, Vicksburg, MS.

11. Mudd, T. J. 1971. "Analysis of Pile Foundation," Paper presented at Structures Conference, Lower Mississippi Valley Division, 23-24 Sep 1969, Vicksburg, MS.

12. Ochoa, M., and O'Neill, M. W. 1988 (Jun). "Lateral Pile-Group Interaction Factors for Free-Headed Pile Groups in from Full-Scale Experiments," Miscellaneous Paper GL-88-12, US Army Engineer Waterways Experiment Station, Vicksburg, MS.

13. Radhakrishnan, N., and Parker, F. 1975 (May). "Background Theory and Documentation of Five University of Texas Soil-Structure Interaction Computer Programs," Miscellaneous Paper K-75-2, US Army Engineer Waterways Experiment Station, Vicksburg, MS.

14. Reese, L. C., Cooley, L. A., and Radhakrishnan, N. 1984 (Apr). "Laterally Loaded Piles and Computer Program COM624G," Technical Report K-84-2, US Army Engineer Waterways Experiment Station, Vicksburg, MS.

15. Smith, William G., and Mlakar, Paul F. 1987 (Jun). "Lumped Parameter Seismic Analysis of Pile Foundations," Report No. J650-87-008/2495, Vicksburg, MS.

16. Strom, Ralph, Abraham, Kevin, and Jones, H. Wayne. 1990 (Apr). "User's Guide: Pile Group - Concrete Pile Analysis Program (CPGC) Computer Program," Instruction Report ITL-90-2, US Army Engineer Waterways Experiment Station, Vicksburg, MS.

17. Tucker, L. M., and Briaud, J. 1988. "Axial Response of Three Vibratory and Three Impact Driven H-Pile in Sand," Miscellaneous Paper GL-88-28, US Army Engineer Waterways Experiment Station, Vicksburg, MS.

18. US Army Engineer Waterways Experiment Station. 1987 (May). "User's Guide: Concrete Strength Investigation and Design (CASTR) in Accordance with ACI 318-83," Instruction Report ITL-87-2, Vicksburg, MS.

B-2. Other Publications. All related publications are available on interlibrary loan from the Research Library, US Army Engineer Waterways Experiment Station, Attn: CEWES-IM-MI-R, 3909 Halls Ferry Road, Vicksburg, MS 39180-6199.

19. American Concrete Institute. 1983. "Building Code Requirements for Reinforced Concrete," ACI 318-83, Detroit, MI.

20. American Concrete Institute. 1986. "Recommendations for Design, Manufacture and Installation of Concrete Piles," ACI 543R-74, Detroit, MI.

21. American Institute of Steel Construction. 1989. Manual of Steel Construction, 9th ed., New York.

22. American Society for Testing and Materials. 1974. "Method for Establishing Design Stresses for Round Timber Piles," D2899-74, Vol 04.09, Philadelphia, PA.

23. American Society for Testing and Materials. 1978. "Method of Testing Individual Piles Under Static Axial Tensile Load," D3689-78, Vol 04.08, Philadelphia, PA.

24. American Society for Testing and Materials. 1981. "Method of Testing Piles Under Lateral Loads," D3966-81, Vol 04.08, Philadelphia, PA.

25. American Society for Testing and Materials. 1983. "Method of Testing Piles under Static Axial Compressive Load," D1143-81, 1986 Annual Book of ASTM Standards, Vol 04.08, Philadelphia, PA.

26. Aschenbrenner, T. B., and Olson, R. E. 1984 (Oct). "Prediction of Settlement of Single Piles in Clay," Analysis and Design of Pile Foundations, American Society of Civil Engineers, J. R. Meyer, Ed.

27. Bowles Foundation Analysis and Design. 1977. 2nd ed., McGraw-Hill, New York.

28. Coyle, H. M., and Reese, L. C. 1966 (Mar). "Load-Transfer for Axially Loaded Piles in Clay," Journal, Soil Mechanics and Foundations Division, American Society of Civil Engineers, Vol 92, No. SM2, pp 1-26.

29. Coyle, H. M., and Sulaiman, I. H. 1967 (Nov). "Skin Friction for Steel Piles in Sand," Journal, Soil Mechanics and Foundations Division, American Society of Civil Engineers, Vol 93, No. SM-6, pp 261-278.

30. Davisson, M. T. 1970. "Lateral Load Capacity of Piles," Highway Research Record No. 333, Highway Research Board, National Academy of Sciences--National Research Council, Washington, DC.

31. Deep Foundations Institute. 1979. "A Pile Inspectors Guide to Hammers," Equipment Applications Committee, Springfield, NJ.

32. Deep Foundations Institute. 1981. "Glossary of Foundation Terms," Equipment Applications Committee, Springfield, NJ.

33. Department of the Navy. 1982 (May). "Foundations and Earth Structures." NAVFAC DM-7.2, Naval Facilities Engineering Command, 200 Stovall St., Alexandria, VA.

34. Federal Highway Administration. 1985. "Manual on Design and Construction of Driven Pile Foundations," Demonstration Projects Division and Construction and Maintenance Division, Washington, DC.

35. Garlanger, J. H. 1973. "Prediction of the Downdrag Load at Culter Circle Bridge," Symposium on Downdrag of Piles, Massachusetts Institute of Technology.

36. Hetenyi, M. 1946. Beams on Elastic Foundation, The University of Michigan Press, Ann Arbor, MI.

37. Hrennikoff, A. 1950. "Analysis of Pile Foundation with Batter Piles," Transactions, American Society of Civil Engineers, Vol 115, No. 2401, pp 351-383.

38. Kraft, L. M., Focht, J. A., and Amerasinghe, S. F. 1981 (Nov). "Friction Capacity of Piles Driven Into Clay," Journal, Geotechnical Engineering Division, American Society of Civil Engineers, Vol 107, No. GT11, pp 1521-1541.

39. Matlock, H. 1970. "Correlations for Design of Laterally Loaded Piles in Soft Clay," Paper No. OTC 1204, Proceedings, Second Annual Offshore Technology Conference, Houston, TX, Vol 1, pp 577-594.

40. Meyerhof, G. G. 1976 (Mar). "Bearing Capacity and Settlement of Pile Foundations," Journal, Geotechnical Engineering Division, American Society of Civil Engineers, Vol 102, No. GT3, pp 197-228.

41. Nathan, Noel D. 1983 (Mar/Apr). "Slenderness of Prestressed Concrete Columns," PCI Journal, Vol. 28, No.2, pp 50-77.

42. Novak, M. 1984 (Nov). "Dynamic Stiffness and Damping of Piles," Canadian Geotechnical Journal, Vol 11, No. 4, pp 574-598.

43. Novak, M., and Grigg, R. F. 1976 (Nov). "Dynamic Experiments With Small Pile Foundations," Canadian Geotechnical Journal, Vol 13, No. 4, pp 372-385.

44. O'Neill, M. W. 1964. "Determination of the Pile-Head, Torque-Twist Relationship for a Circular Pile Embedded in a Clay Soil," M.S. thesis, University of Texas, Austin, TX.

45. O'Neill, M. W., and Tsai, C. N. 1984 (Nov). "An Investigation of Soil Nonlinearity and Pile-Soil-Pile Interaction in Pile Group Analysis," Research Report No. UHUC 84-9, Department of Civil Engineering, University of Houston, prepared for US Army Engineer Waterways Experiment Station, Vicksburg, MS.

46. Peck, R. B., Hanson, W. E., and Thornburn, T. H. 1974. Foundation Engineering, Wiley, New York.

47. PCI Committee on Prestressed Concrete Columns. 1988 (Jul-Aug). "Recommended Practice for the Design of Prestressed Concrete Columns and Walls," Vol 33, No. 4, pp 56-95.

48. Poulas, H. G., and Davis, E. H. 1980. Pile Foundation Analysis and Design, Wiley, New York.

49. Prakash, S. 1981. Soil Dynamics, McGraw-Hill, New York.

50. Reese, L. C. 1975 (Mar). "Laterally Loaded Piles," Design, Construction, and Performance of Deep Foundations Lecture Series, University of California, Berkeley.

51. Reese, L. C., Cox, W. R., and Koop, F. D. 1974. "Analysis of Laterally Loaded Piles in Sand," Paper No. OTC 2090, Proceedings, Sixth Offshore Technology Conference, Houston, TX, Vol 2, pp 473-483.

52. Reese, L. C., Cox, W. R., and Koop, F. D. 1975. "Field Testing and Analysis of Laterally Loaded Piles in Stiff Clay," Paper No. OTC 2312, Proceedings, Seventh Offshore Technology Conference, Houston, TX, pp 671-690.

53. Reese, L. C., and Welsh, R. C. 1975 (Jul). "Lateral Loading of Deep Foundations in Stiff Clay," Journal, Geotechnical Engineering Division, American Society of Civil Engineers, Vol 101, No. GT7, pp 633-649.

54. Saul, William E. 1968 (May). "Static and Dynamic Analysis of Pile Foundations," Journal, Structural Division, American Society of Civil Engineers, Vol 94, No. ST5.

55. Scott, F. S. 1981. Foundation Analysis, Prentice-Hall, Englewood Cliffs, NJ.

56. Semple, R. M., and Rigden, W. J. 1984 (Oct). "Shaft Capacity of Driven Pile in Clay," Analysis and Design of Pile Foundations, American Society of Civil Engineers, J. R. Meyer, Ed., pp 59-79.

57. Stoll, U. W. 1972 (Apr). "Torque Shear Test of Cylindrical Friction Piles," Civil Engineering, American Society of Civil Engineers, Vol 42, pp 63-65.

58. Teng, W. C. 1962. Foundation Design, Prentice-Hall, Englewood Cliffs, NJ.

59. Terzaghi, Z., and Peck, R. B. 1967. Soil Mechanics in Engineering Practice, Wiley, New York.

60. Vesic, A. S. 1977. "Design of Pile Foundations," National Cooperative Highway Research Program, Synthesis of Highway Practice No. 42, Transportation Research Board, Washington, DC.

61. Vijayrergiya, V. N. 1977 (Mar). "Load - Movement Characteristics of Piles," Ports '77, Proceedings, 4th Annual Symposium of the Waterways, Port, Coastal and Ocean Division of American Society of Civil Engineers, Vol 2, pp 269-284.

APPENDIX C

CASE HISTORY

PILE DRIVING AT LOCK AND DAM NO. 1
RED RIVER WATERWAY

CASE HISTORY

PILE DRIVING AT LOCK AND DAM No. 1
RED RIVER WATERWAY

I. Introduction.

    This case history report has been prepared to present an overview of
problems associated with the driving of steel H-piles at the Red River Lock
and Dam No. 1 in the New Orleans District of the Lower Mississippi Valley
Division.

    The discussion of problems encountered such as pile interference, piles
hitting early refusal, driving with a vibratory hammer, etc., and the solu-
tions and recommended preventive measures for these problems are presented for
others who may encounter similar situations and benefit from this experience.

II. Description of Project.

    Lock and Dam No. 1 is a feature of the Red River Waterway Project (Mis-
sissippi River to Shreveport), located in Catahoula Parish, Louisiana,
approximately 5 miles north of the town of Brouillette. (See Figure C-1.)
Navigation from the Mississippi River to the Red River is now provided through
Old River via Old River Lock. The purpose of this project is to provide for
navigation on the Red River from its junction at Old River up to Shreveport,
Louisiana.

    The Lock is a soil-supported reinforced concrete U-frame structure with a
useable length of 685 ft and a width of 84 feet. The dam is a reinforced
concrete structure with eleven 50-foot wide steel tainter gates. (See
Figure C-2.) Both the dam structure and its stilling basin are supported on
steel H-piles.

III. Phases of Construction.

    To expedite the construction schedule for Lock and Dam No. 1 phased con-
struction was necessary. A contract for the initial excavation and construction of the earthen cofferdam was awarded in June 1977 (Phase I). A second
contract, which included installation of the dewatering system, structural
excavation and construction, driving and testing piles, and driving of the dam
service piles, was awarded in July 1978 (Phase II).

    Although the specifications properly stated that the actual pile lengths
would be determined from the test pile driving and loading, because of time
constraints, production piles were ordered and purchased to the lengths as
determined from soil parameters obtained from the soil borings. Numerous
delays encountered by the Contractor in starting the pile driving led to the
deletion of the driving of the service piles from this contract.

    The third contract was awarded in December 1979. It consisted of all the
remaining work and included the driving of the service piles which had to be
incorporated into the final plans and specifications for this phase of construction (Phase III) due to its earlier deletion from Phase II.

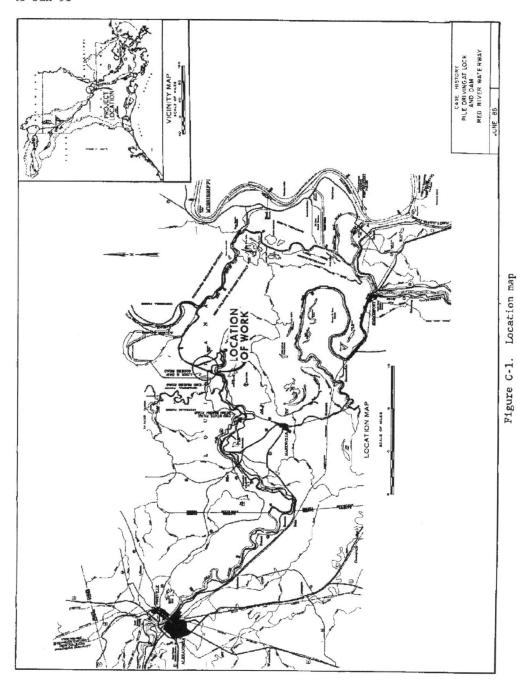

Figure C-1. Location map

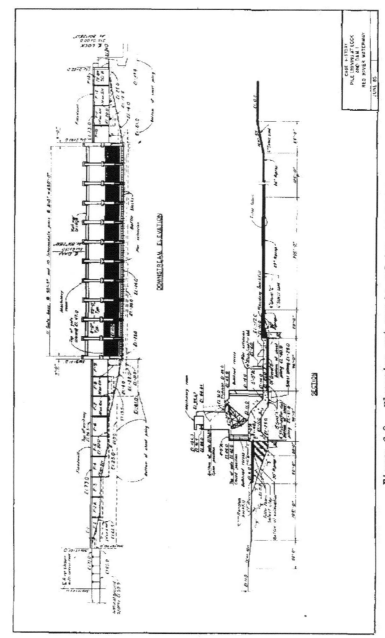

Figure C-2. Elevation and section of dam structure

C-5

IV. Soil Conditions.

      A layer of medium to stiff holocene backswamp clays is located along the
dam axis extending from the base of the dam (approximately -11 N.G.V.D.) to
the top of the sand stratum (approximately elevation -48 N.G.V.D.). The bottom of this sand stratum extends to below elevation -140 N.G.V.D.

V. Driving Hammers.

      The Phase II pile specifications allowed for the use of a vibratory
hammer which the Contractor elected to use. Pile tests were to be performed
on HP 14x89 and HP 12x53 steel H-piles. The Contractor drove the test piles
using a Foster Vibratory Hammer Model 4000, having an eccentric moment of
4,000 inch-pounds and a weight of 18,800 pounds. Using this hammer, the 60 to
80 foot HP 12x53 test piling were driven in less than 2 minutes, while the
87 to 106 foot long HP 14x89 piling were driven in 10-16 minutes driving time.
The various test piles were instrumented with "telltales" to determine the
pile capacity in the various strata. Only load carrying capacity in the sand
substratum was used in the design since it was determined that the long-term
support capabilities of the overburden clays and silts would be negligible.
The total pile support in the soil was to be developed by the skin friction in
the sand and the pile tip bearing.

      All compression pile tests were conducted 7 to 8 days after driving with
tension pile tests conducted 11 to 15 days after driving. Two compression
tests were conducted at site PT-1S on HP 14x89 piles 1-C and 2-C. The results
of these tests indicated pile capacities of 63 and 83% of the computed theoretical capacity, respectively. At site PT-A, for 4 HP 12x53 piles (1-C, 2-C,
3-C, and 2C-X) tested in compression, the results were 37, 25, 40 and 57% of
the computed theoretical capacities, respectively. In tension, the HP 12x53
pile 2C-X developed 30% of the computed capacity.

      Retesting of some of the piles was directed to determine what effects
elapsed time would have on the capacities of the piling. A compression re-
test, 25 days after driving, was made on HP 14x89 pile 1-C which originally
tested at 63% of the computed capacity. The capacity of the pile increased
from 63% to 78% of the computed capacity indicating a strength growth with
elapsed time. The compression retests of two of the HP 12x53 piles resulted
in increases from 37% to 70% of the computed capacity for pile 1-C, retested
70 days after driving; and from 40% to 78% for pile 3-C, retested 56 days
after driving. Two tension pile retests were made, one test after 22 days had
elapsed from the time of driving and the other test performed after 54 days
had elapsed. An increase from 41% to 86% in HP 14x89 pile 2-C and an increase
from 30% to 67% in HP 12x53 pile 2C-X of the computed capacity resulted for
these two tension pile retests.

      Consequently, the retests for both compression and tension test piles
resulted in increased pile capacities with the passage of time for piling
driven with a vibratory hammer. Figure C-3 is a table showing the capacities
and dates driven for the tests and retests in compression and tension.

      As a result of the pile tests and retests the service piling had to be
approximately 12 feet longer than the computed theoretical lengths. As stated
previously, the steel H-piling had been ordered and delivered prior to the
pile tests, therefore splicing to obtain the additional length piling was

| | | | PILE DATA | | | | COMPUTED COMP. LOAD(TON)* | COMPRESSION | | | COMPRESSION | | | COMPUTED TENSION LOAD(TON) | TENSION | | | TENSION | |
| | | | | | | | | TEST | | | RE-TEST | | | | TEST | | | RE-TEST | |
SITE	PILE NO.	PILE TYPE	BUTT ELEV	TIP ELEV	L	DATE DRIVEN		DATE	LOAD*	%**	DATE	LOAD*	%**		DATE	LOAD*	%**	DATE	LOAD*	%**
PT-IS	1-C	14x89	-18.6'	-105.2	86.6'	7/9/79	176^T	7/17/79	110^T	63	8/3/79	137^T	78							
PT-IS	2-C	14x89	-18.7	-124.2	105.5	7/9/79	228^T	7/16/79	190^T	83				126^T	7/20/79	51^T	41	7/31/79	108^T	86
PT-A	1-C	12x53	-20.1	-80.3	60.2'	7/11/79	87^T	7/19/79	32^T	37	9/19/79	60^T	70							
PT-A	2-C'	12x53	-20.8	-89.3	68.5'	7/11/79	110^T	7/18/79	26^T	25										
PT-A	3-C	12x53	-20.8	-89.0	68.2'	8/1/79	110^T	8/9/79	44^T	40	9/27/79	86^T	78							
PT-A	2C-X	12x53	-20.7	-99.2	78.5'	8/1/79	134^T	8/8/79	76^T	57				69^T	8/16/79	21^T	30	9/25/79	46^T	67

*LOAD IN SAND.

**PERCENTAGE OF COMPUTED LOAD

Figure C-3. Red River Lock 8 Dam No. 1 pile test summary

necessary. The required splicing was performed in the storage area during the
Phase III contract, increasing the maximum pile length from 96 feet to
108 feet.

The pile splice detail required a full penetration butt weld with
3/8-inch thick fish plates over the outside of the flanges (Figure C-4).

VI. Pile Layout.

The pile layout for the dam foundation consisted of alternating rows of
batter piles, with a downstream batter of 2.5 vertical to 1 horizontal and an
upstream batter of 4 vertical to 1 horizontal. (See Figure C-5 for typical
layout.) Typical pile spacing between rows was 5 feet center to center.
There were 904 HP 14x89 piles and 1,472 HP 14x73 piles in the dam foundation.
The stilling basin, which consisted of predominantly 60-foot vertical piling,
had 388 HP 14x73 piles and 633 HP 12x53 piles. (See Figure C-6 for typical
layout.)

VII. Pile Driving Specifications.

The pile specifications allowed for a driving tolerance of 1/4 inch per
linear foot along the longitudinal axis of the pile. A 3-inch tolerance in
both X and Y directions in the position of the butt of the pile was allowed.
The specifications also required that the pile be sufficiently supported in
the leads.

Since the test piling were driven with a vibratory hammer, the specification for the Phase III contract mandated the use of a vibratory hammer with
the same effective driving energy and efficiency. The Contractor elected to
use an ICE (International Construction Equipment, Inc.) model 812 Vibratory
Hammer. This hammer has an eccentric moment of 4,000 inch-pounds and weighs
15,600 pounds.

The Contractor's driving equipment included two 4100 Manitowac Cranes
fitted with International Construction Equipment (ICE) fixed leads and hydraulic spotters. An intermediate support was not provided in the leads, consequently the pile was supported only at the hammer and the ground, and was free
to move between these two points.

VIII. Pile Driving.

Within the first few days of driving, several piles were reported to have
reached refusal above final grade. The specifications defined refusal as a
penetration rate of less than a tenth of a foot per minute. The piles reached
refusal at depths ranging from 37 feet to 85 feet as measured along the length
of the piling. Some of the piling that refused above the specified grade were
ordered pulled. Upon examination of these piles it was apparent from the
bottom damage that there had been pile interference during driving and colli-
sion with other piles.

A review of the Contractor's driving procedures revealed the following:

a. The batter was set by using a 5-foot carpenter's level with a
template. The template, which was a triangle set to the prescribed batter,

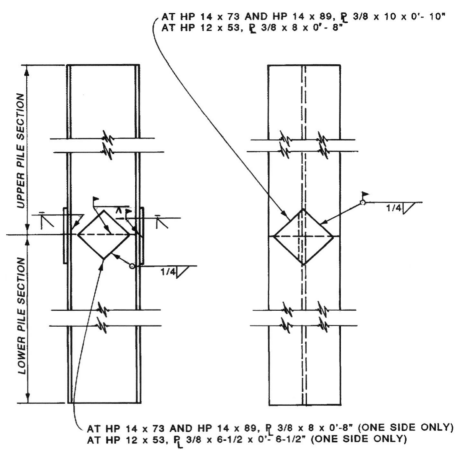

AT HP 14 x 73 AND HP 14 x 89, ℄ 3/8 x 10 x 0'- 10"
AT HP 12 x 53, ℄ 3/8 x 8 x 0'- 8"

1/4

UPPER PILE SECTION

LOWER PILE SECTION

1/4

AT HP 14 x 73 AND HP 14 x 89, ℄ 3/8 x 8 x 0'-8" (ONE SIDE ONLY)
AT HP 12 x 53, ℄ 3/8 x 6-1/2 x 0'- 6-1/2" (ONE SIDE ONLY)

NOTE: SPLICE LAYOUT IS TYPICAL
FOR HP 14 x 73, HP 14 x 89
AND HP 12 x 53

Figure C-4.   H-pile splice layout

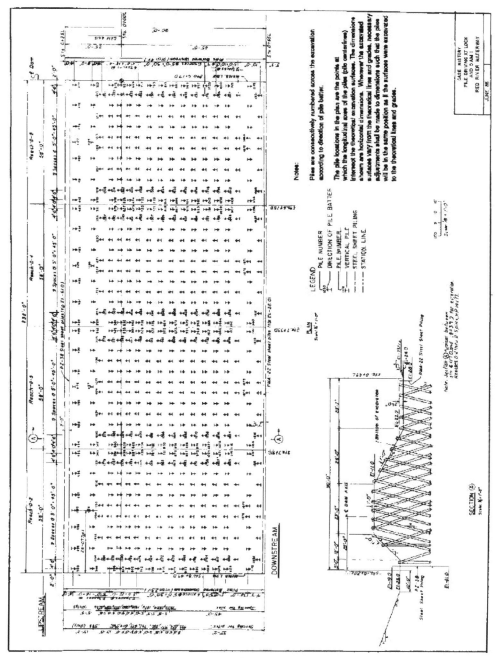

Figure C-5. Typical dam monoliths-pile layout

C-10

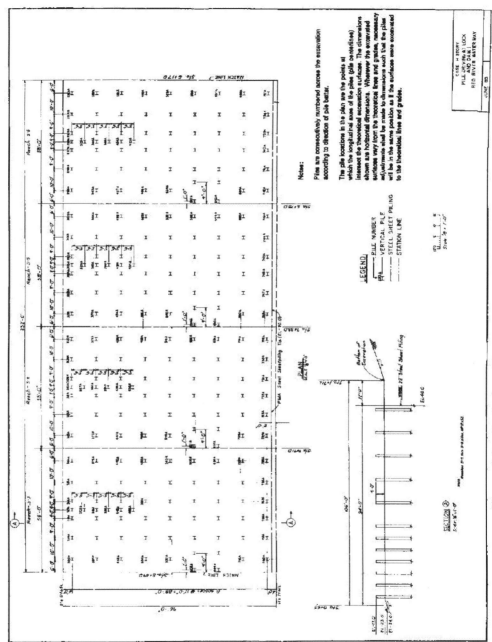

Figure C-6. Typical stilling basin monoliths-pile layout

was placed against the leads, with a level used to plumb the vertical leg of the triangle.

       b. Alignment of the pile along a line perpendicular to the axis of the dam was being accomplished by "eye-balling" the leads. Since a slight error in alignment would lead to pile interference, this method of alignment was not acceptable.

       c. The Contractor was observed to be picking piles up at their ends in lieu of the specified one- and two-point pick up locations shown on the drawings. This method of pick up was causing excessive sweep in the piles.

       d. Piling were being dragged one pile at a time with a front end loader from the storage area to the driving area. The lifted end of the pile was approximately 4 feet above the ground. This method of transport, along with improper storing, was causing damage to the piles. Permanent sweep of magnitudes 1-1/2 to 2 feet was observed on some of the piles.

       e. During driving, the pilings were supported at the hammer and at the ground with no intermediate support point in the leads. Therefore, the piles which were 90-108 feet long had a tendency to buckle during driving. This buckling had the effect of increasing alignment problems.

It was concluded that the driving difficulties were caused by pile interference. Pile interference resulted from piles being driven off-line and/or driving of piles having substantial sweep caused by improper handling. The Contractor was directed to take the following remedial measures:

       a. Use a transit to set the alignment of the piling perpendicular to the axis of the dam. Set up a transit on the pile-driving line, and by shooting the top and bottom of the leads, bring the pile into alignment.

       b. Prior to driving, straighten all piles having more than 6" of sweep.

       c. Revise all pile handling operations utilizing the one- and two-point pick ups shown on the plans.

       d. Handle the piles such that deflections and/or bending stresses are resisted by the strong axis of the pile.

       e. Add an additional pile support in the leads to support the piles at their midpoints.

IX. <u>Piles Hitting Refusal</u>.

Within the next several weeks after the above remedial measures were put into effect, additional piles hit refusal during driving. The pattern of refusal was very erratic. Piles adjacent to refused piles could possibly be driven to grade with no problem. On some occasions, refused piles could be pulled and redriven to grade within 1.5 feet of the original location. Average time of driving was approximately 7 minutes.

About this time several theories were developed as to the reasons for pile refusals; 1) the piles were refusing due to obstructions in the foundation; 2) the vibratory hammer was delivering insufficient energy to the pile tip; 3) or a combination of the above. Because of all these uncertainties, William A. Loftus, an expert in the field of pile driving was brought in by the New Orleans District for advice on the problem of pile refusals. After visiting the site and observing the Contractor's pile driving operations, Mr. Loftus made the following observations and recommendations:

      a. The corrective measures that were placed upon the Contractor had resulted in adequate control of pile alignment and placement. The possibility that sand densifications and obstructions in the sand stratum, such as gravel layers were more likely than had been expected from the pile tests and borings could be the cause of early pile refusal.

      b. A load test of an in-place pile that reached early refusal should be made to determine the capacity.

      c. Large diameter borings should be taken to determine the extent and size of gravel which was suspected as a cause of pile refusal.

The above recommendations were adopted, and six large-diameter borings were taken using a 3-inch spiltspoon and a 6-inch core barrel. Gravel zones were found to exist below elevation -85. The average size of gravel was 3/4 to 1-1/2 inches. Gravels larger than 2-1/2 to 3 inches were uncommon; however, cobbles as large as 4 inches were encountered. The original design borings were small diameter borings and could not detect these cobbles.

About 15 percent of the 2,376 piles in the dam foundation (excluding the shorter stilling basin piles which were not a problem) reached refusal above grade. Slightly less than 10 percent of the piles were cut off. Typically, about 20 feet of piling was cut off with the maximum being 40 feet. Shoes were used on about 10 percent of the piles. Compression and tension pile load tests were performed on a service pile that was cut off 12 feet above grade and a service pile that was driven to grade, and an evaluation of the as-built pile foundation with actual pile lengths was accomplished. This evaluation concluded that the pile foundation was safe and provided an adequate degree of stability for the dam monoliths.

X. Conclusions.

In reviewing the history of the pile tests, the driving procedure, the design, and the specifications for the Phase II and Phase III contracts for Lock and Dam No. 1, there are several factors that led to the many problems encountered during the pile driving operations. A sample pile specification which attempts to remedy these problems, is included as an attachment. This attachment was developed solely for this particular case history. Future projects should utilize the guide specification with appropriate modifications as necessary.

A. Factors Leading to Problems:

The vibratory hammer was very efficient in driving the piles to grade, but apparently reduced the capacity of the test piles. The phenomenon which causes this apparent loss of capacity is unknown, but there was sufficient evidence in the retests that showed a growth of pile strength with time.

Secondly, the Contractor started driving piling with very poor alignment and quality control procedures. This undoubtedly led to pile interference problems.

Thirdly, there existed cobbles in the foundation which were undetected by the original small diameter bore holes. These cobbles were the cause of early pile refusals after the Contractor's pile driving procedures were improved.

B. Vibratory Hammer:

The vibratory hammer is a very efficient device for driving piles and therefore it's use reduces pile driving costs. In addition to the driving economics of the vibratory hammer in a sand foundation, the hammer has other good attributes. These attributes include reduced noise, alignment control since the hammer grips the pile at its butt, and easier pile extraction since the hammer can be used without rerigging. Consequently, we should not indiscrimently prohibit its use. However, until further data are obtained for the vibratory hammer relative to reduced pile capacity, it is recommended that specifications allow the Contractor the option to use a vibratory hammer or a combination of the vibratory hammer with an impact hammer used for seating the piles with the following qualifications:

1. The Contractor perform duplicate pile tests for the vibratory hammer at his expense.

2. If increased pile lengths are required because of the vibratory hammer, the Contractor should bear the cost of the added lengths (specifications for the Old River Auxiliary Structures contained these stipulations).

C. Handling and Driving Piles:

It is extremely important that proper care should be taken in setting batters and aligning the piles. This is particularly important in congested pile foundations where a slight misalignment may cause pile interference. The specifications should require piles to be driven within a tolerance of 1/4 inch per linear foot of piling.

1. Piles with excessive sweep should be straightened before driving.

2. Intermediate supports in the leads should be provided to restrain piles from buckling during driving.

3. Care should be taken during driving, storing, handling, and transporting piles to avoid damage to the piling.

D. Sufficient Soils Data:

It seems that in the design of any foundation, there can never be too much information on the stratigraphy. A pile foundation is no exception. Where there is evidence of gravel layers or cobbles, it is recommended that large-diameter borings be taken to determine the extent and size of the gravel or cobbles.

E. Pile Refusal:

Refusal of piling in most instances is an indication of extreme point bearing capacity. Therefore, unless pile interference or a malfunctioning hammer system is suspected and the piling is not an uplift (tension) pile, it may be acceptable to cutoff the pile. For the refusal of tension piles, consideration has to be given to the number of piles that refuse, the required capacity and what reduction in safety factors can be accepted. If a large percentage of piles reach refusal, then an in situ pile test or changes in pile driving, e.g., hammer type of size, or use of pile shoes may be necessary. However, since construction delays are inherent in testing and the associated costs to the Contractor and to the Government are very high, this procedure should be used only when absolutely necessary.

F. Layout of Pile Foundation:

In the structural design and layout of the pile foundation, piles should be adequately spaced to allow for some pile drift. Although good pile alignment and batter can be achieved in the leads, it is virtually impossible to control or detect pile drift in the ground. It is recommended that the specified tolerance of 1/4 inch per linear foot of pile be used to determine the minimum pile spacing. Thus, if piling were 100 feet long, a minimum clear spacing of 50 inches would be provided to allow for the tolerance in each of the piles (2 x 100' x 1/4").

ATTACHMENT

SECTION 2I - STEEL H-PILES

PART 1 - GENERAL

1. SCOPE. The work covered by this section consists of furnishing all plant, equipment, labor, and materials and performing all operations in connection with the installation of steel H-piles in accordance with these specifications and applicable drawings. Tension anchors are covered under SECTION 5J. The pile load and driving test program is covered in SECTION 2K.

2. QUALITY CONTROL.

2.1 General. The Contractor shall establish and maintain quality control for all operations to assure compliance with contract requirements and maintain records of his quality control for all construction operations, including but not limited to the following:

> (1) Material
> (2) Storing and handling
> (3) Placing (location, alignment, etc.)
> (4) Driving and splicing
> (5) Cutting

2.2 Reporting. The original and two copies of these records and tests, as well as corrective action taken, shall be furnished to the Government daily. Format of the report shall be as prescribed in SP-16.

3. APPLICABLE PUBLICATIONS. The following publications of the issues listed below, but referred to thereafter by basic designation only, form a part of this specification to the extent indicated by the references thereto:

3.1 American Society for Testing and Materials (ASTM) Standards.

A 36-77a   Structural Steel

3.2 American Institute of Steel Construction (AISC).

Manual of Steel Construction, 9th Edition

4. SUBMITTALS.

4.1 Equipment Descriptions. The Contractor shall submit complete descriptions of pile driving equipment, including hammers, extractors, and other appurtenances to the Contracting Officer for approval 45 days prior to commencement of work.

4.2 Handling and Storage Plans. Proposed methods of handling and stockpiling the piles shall comply with requirements of 2I-6.4 and 2I-6.5 and shall be submitted in detail to the Contracting Officer for review and approval at least 45 days prior to delivery of the production piles to the job site.

4.3 Control of Placement. Proposed methods of controlling the location and alinement of the production piles shall comply with requirements of 2I-6.6

and shall be submitted in detail to the Contracting Officer for review and approval at least 45 days prior to driving the first production pile. Description of the alinement controls shall include the proposed methods of controlling the pile batter, the vertical plumbness, and rotation of the pile about the longitudinal centerline of the web.

4.4 Driving Record. The Contractor shall furnish daily to the Contracting Officer a copy of all driving data required in 2I-6.7.1. Unusual driving conditions, interruptions or delays during driving, and any other information associated with the pile driving operations shall be noted. The records shall be submitted in triplicate (original and two copies).

PART 2 - PRODUCTS

5. MATERIALS.

5.1 Steel. Steel for H-piles and splice plates shall conform to the requirements of ASTM A 36.

5.2 Steel H-Piles. Steel H-piles shall be of the shape and sections shown on the drawings. Piles shall have standard square ends, unless otherwise specified or directed. Lengths of piles shall be determined as specified below in 2I-6.3.1.

5.3 Pile Points. Pile points shall be Pruyn H-Pile Points, Type BP-75750, or approved equal.

5.4 Steel H-Pile Splices. Pile splice plates shall conform to details shown on the drawings. All welding shall be performed by certified welders as specified in SECTION 5E. The Government will test select splices by nondestructive methods.

5.5 Pile Tension Anchors. Pile tension anchors shall conform to details shown on the drawings. The Government will test the capacity of select tension anchors.

PART 3 - EXECUTION

6. INSTALLATION.

6.1 Pile Driving Equipment. Pile driving hammers shall be steam, air or diesel operated impact, single-acting, double-acting or differential acting type. Vibratory hammers will be allowed provided applicable requirements of SECTION 2K are satisfied. The production piles shall be driven with the same size and type hammer, operating with the same effective energy and efficiency as used in driving the steel test piles (which are covered in SECTION 2K). All pile driving equipment and appurtenant items shall be equal to that used in the test pile driving operations. The size or capacity of hammers shall be as recommended by the manufacturer for the pile weights and soil formation to be penetrated. Impact hammers shall have a minimum energy of 24,000 ft-lb for 14-inch H-piles and 19,500 ft-lb for 12-inch H-piles. The hammers shall be operated at all times at the speed and conditions recommended by the manufacturer. Boiler, compressor, or engine capacity shall be sufficient to operate the hammer continuously at full rated speed and inlet pressures. Once the actual driving has begun, all conditions (such as alinement, batter, cushions, etc.) shall be kept as constant as possible. Hammers shall have firmly supported leads extending to the lowest point the hammer must reach. In order to reduce the unbraced length of the pile during driving, the Contractor shall provide intermediate support for the pile in the leads at no additional cost to the Government. The Contractor shall submit details on intermediate supports to the Contracting Officer for approval.

6.2 Test Piles. All work associated with the required pile tests is covered in SECTION 2K.

6.3 Permanent Production Piles.

6.3.1 Lengths. The estimated quantities of piles listed in the unit price schedule are given for bidding purposes only. The Contractor shall not order piling until he receives a quantities list from the Contracting Officer as specified below. The Contracting Officer will determine the actual lengths of production piles required to be driven below cut-off elevation for the various locations in the work and will furnish the Contractor a quantities list which indicates lengths and locations of all piles to be placed. This determination will be made from the results of the pile driving and load tests performed as specified in SECTION 2K of this specification and will be pro- vided to the Contractor as specified in SECTION 2K. Piling shall be furnished full length. Splicing of piling to make up the required lengths will not be permitted. The Contractor's schedule shall be established to assure timely accomplishment of the pile tests, an essential item along the critical path.

6.4 Storing. Steel H-piles stored at the job site shall be stored on a level surface in an area that will not pond water and the piles shall be stacked in such a manner that all piles have uniform support along their length without sagging or bending. If it is not feasible to store the piles on a hard level surface, hardwood blocking shall be laid in such a manner so that piles are brought to level. Blocking shall be spaced at distances suf- ficiently short to prevent sagging or bending. In no case shall blocking be more than 10 feet apart nor more than 2 feet from the ends of the pile. The method of stacking shall not result in damage to the pile or excessive sweep or camber. Plan for storing H-piles shall be submitted as specified in 2I-4.2.

6.5 Handling. Pick-up points for steel H-piles shall be as shown on the drawings and shall be plainly marked on all piles. All lifting shall be done at these points. All lifting, except for lifting the pile into the driving leads, shall be accomplished using a two point pick-up. A one point pick-up may be used for lifting the pile into the driving leads. Pick-up devices shall be of the type that clamp to both pile flanges at each pick-up point. Use of alternate types of pick-up devices shall be subject to approval by the Contracting Officer. Burning holes in flanges or webs for handling shall not be permitted. During on-site transporting of piles, the piles shall be main- tained in a straight position and shall be supported, as a minimum, at the quarter points. Dragging of piles across the ground shall not be permitted. Before the piles are transported from the stockpile area to the driving area, all piles shall be inspected for damage and excessive sweep and camber in accordance with these specifications and the drawings. The web and flanges of the piles shall be checked by rotating the pile with the pile resting on a firm level surface. A pile which has camber and/or sweep greater than 2 inches shall be rejected and shall not be transported to the driving areas.

A pile which is damaged and which in the opinion of the Contracting Officer is unusable, will be rejected, and shall not be transported to the driving areas. After the piles are delivered to the driving area, they shall be checked again, visually, to insure that damage has not occurred during handling and transporting from the stockpile area to the driving area. Any pile which is damaged and which damage, in the opinion of the Contracting Officer, renders the pile unusable and/or which contains excessive sweep or camber, as defined above, shall be replaced by a new pile at no additional cost to the Government. Proposed methods of handling shall be submitted as specified in 2I-4.2.

6.6 <u>Placement</u>. Piles shall be accurately placed in the correct location and alinements, both laterally and longitudinally and to the vertical or batter lines as shown on the contract drawings. To insure correct placement of piles, the Contractor shall establish a rectangular grid system by use of a surveying instrument having at least the accuracy of a one-second Theodolite. The Contractor shall check each pile, prior to driving and with the pile head seated in the hammer, for correct batter, vertical plumbness, and rotation of the pile about the centerline of the web. The vertical plumbness of the pile shall be checked with a surveying instrument having at least the accuracy of a one-second Theodolite to insure that the pile is being driven parallel to the grid line in the direction of the pile batter. A lateral deviation from <u>the correct location at</u> the cut-off elevation of not more <u>than 3 inches will be</u> permitted. A vertical deviation from the correct cut-off elevations shown on the drawings of not more <u>than +2 inches will</u> be permitted. A variation in alinement of not more <u>than 1/4-inch per foot of longitudinal axis will be permitted</u>. Moving the pile by rotating the leads, pulling on the pile, and wedging the pile will not be allowed. If, during driving, the pile shifts or otherwise moves beyond the specified tolerances, the Contractor may be required to pull and redrive the pile as directed by the Contracting Officer. Piles which are misplaced shall be pulled and redriven at no additional cost to the Government. Any voids that remain after pulling misplaced piles shall be filled with sand, and all costs related thereto shall be borne by the Contractor.

6.7 <u>Pile Driving</u>.

6.7.1 <u>Driving</u>. No piling shall be driven <u>within 100 feet of any concrete structure, unless authorized</u> by the Contracting Officer. A complete and accurate record of the driving of piles as specified in 2I-4.4 shall be compiled by the Contractor for submission to the Contracting Officer. This record shall include pile dimensions and locations, the description of hammer used, rate of hammer operation, for impact hammers the number of blows required for each foot of penetration throughout the entire length of each pile, for vibratory hammers the cumulative time of penetration at five foot intervals shall be recorded throughout the entire length of each pile, butt elevation upon completion of driving and any other pertinent information requested by the Contracting Officer. When driving long piles of high slenderness-ratio, special precautions shall be taken to insure against overstressing and leading away from a plumb or true position. During driving, pile driving hammers shall be operated at all times at the speed, inlet pressure, and conditions recommended by the hammer manufacturer. Each pile shall be driven continuously and without interruption to the minimum required depth of penetration. Deviation from this procedure will be permitted only for cases where interruptions due to splicing as described below or the driving is stopped by causes which reasonably could not have been anticipated. If the pile is

driven with an impact hammer to the minimum depth of penetration but the
minimum penetration per blow has not been attained, the pile shall be driven
deeper as necessary to attain the minimum penetration per blow. The minimum
penetration per blow will be determined by the Contracting Officer upon
completion of the pile tests. A pile which has not reached the minimum
penetration rate per blow when the top has been driven to the cut-off elevation shall be spliced as shown on the drawings and driven to a depth sufficient to develop the minimum penetration rate per blow. All pile splices
shall be fabricated by qualified welders. For impact hammers, when the maximum permissible blows of <u>17 blows per inch for 3 consecutive inches (single-
acting)</u> or <u>20 blows per inch for 3 consecutive inches (double-acting)</u>
is reached above the minimum tip elevation, the pile shall be pulled and redriven
or shall be cut off and either used or abandoned, as directed by the Contracting Officer. For vibratory hammers, when the tip does not move more than
0.1 foot per minute, the Contractor shall immediately attempt driving of the
pile with an impact type hammer conforming to 2I-6.1. If redriving is necessary, piles shall be redriven at a site specified by the Contracting Officer.
Piles which have been uplifted after driving shall be redriven to grade after
conclusion of other driving activity in that general area. if backdriving is
required an equitable adjustment in contract time and price will be made in
accordance with the General Provision Clause "Changes". Jetting shall not be
used to assist driving the piles. Pile points shall be installed only when
directed by the Contracting Officer. Method of installation shall be as
recommended by pile point manufacturer.

6.7.1.1. The Contracting Officer may require that any pile be pulled for
inspection. Any pile which is damaged because of internal defects or by improper handling or driving or is otherwise damaged so as to impair it for its
intended use, shall be removed and replaced. Piles pulled at the direction of
the Contracting Officer and found to be in suitable condition shall be redriven to the required depth at a site specified by the Contracting Officer.
Any pile that cannot be driven to the required depth because of an obstruction
shall be pulled and redriven at a site specified by the Contracting Officer.
Payment for pulled piles will be made in accordance with 2I-8.3.

7. MEASUREMENT.

7.1 <u>Steel H-Piles</u>. Measurement for furnishing and delivering steel
H-piles will be made by the linear foot for steel H-piles acceptably delivered
at the site. Steel H-piles driven with an impact hammer will be measured for
payment on the basis of lengths along the axis of the pile in place below the
cut-off elevation. Steel H-piles driven with a vibratory hammer will be
measured for payment on the basis of pile lengths determined from the testing
program for piles driven with an impact hammer, see SECTION 2K. Pile lengths
will be measured to the nearest tenth of a foot.

8. PAYMENT.

8.1 <u>Furnishing and Delivering Piles</u>. Payment for furnishing and delivering steel H-piles, at the site, will be made at the applicable contract unit
prices for:

"Furnish and Deliver Steel H-Piles (HP 14 x 89)",
"Furnish and Deliver Steel H-Piles (HP 14 x 73)", and
"Furnish and Deliver Steel H-Piles (HP 12 x 53)".

Payment for furnishing and delivering will be made after proper storage of the piling.

8.1.1 Driving Piles. Payment for the measured length of each pile acceptably driven will be made at the applicable contract price per linear foot for "Drive Steel H-Piles (HP 14 x 89)", "Drive Steel H-Piles (HP 14 x 73)", and "Drive Steel H-Piles (HP 12 x 53)". These prices shall include all items incidental to driving the piles and cutting off all piles at the cut-off elevation. Payment for furnishing and installing tension anchors will be made according to SECTION 5J. No additional payment will be made for the use of an impact hammer on piles which have refused with a vibratory hammer. Payment for furnishing and driving test piles will be made according to SECTION 2K.

8.2 Reserved.

8.3 Pulled Piles.

8.3.1 Undamaged Pile. Piles which are pulled at the direction of the Contracting Officer and found to be in good condition will be paid for at the applicable contract unit price for "Furnish and Deliver Steel H-Piles" and "Drive Steel H-Piles" in its original driven position. The cost of pulling and backfilling with sand, if applicable, will be paid for at the applicable contract unit price for "Drive Steel H-Piles". Such pulled piles when re-driven will be paid for at the applicable contract unit price for "Drive Steel H-Piles".

8.3.2 Damaged Pile. Where a pile is pulled and found to be defective and or damaged due to Contractor negligence or internal defects, no payment will be made for either originally furnishing and driving such pile or for the operation of pulling and backfilling with sand, if applicable, and it shall be replaced by a new pile which will be paid for at the applicable contract unit prices. Piles which are pulled and found to be damaged through no fault of the Contractor will be paid for the applicable contract unit price for "Furnish and Deliver Steel H-Piles" and "Drive Steel H-Piles" in its origi-nally driven position. The cost of pulling and backfilling with sand, if applicable, will be paid for at the applicable contract unit price for "Drive Steel H-Pile". A new pile shall be driven in place of the defective and/or damaged pile and will be paid for at the applicable contract unit prices.

8.4 Steel H-Pile Splices. For each pile splice directed by the Contract-ing Officer, payment will be made at the rate of $200 per splice. This price shall include the cost of furnishing all plant, labor and material required to make the directed splices.

8.5 Pile Points. If pile points are required, as directed by the Con-tracting Officer, payment will be made at the rate of $150 per pile point. This price shall include all costs incidental to furnishing and properly in-stalling the pile points on the pile.

APPENDIX D

PILE CAPACITY COMPUTATIONS

D-1. General. After the shear strength and stratification has been selected, the capacity of piles may be computed. The geotechnical engineer computes the capacity of a single pile placed in the subgrade at various levels, then furnishes to the structural engineer a curve relating the pile tip elevation to the axial capacity. These computations may be for any number of pile types, i.e., timber, concrete, steel H-pile, etc., and the computations should be both in the construction shear strength case Q and the long-term shear strength case S. Therefore, two curves for each compression and tension loading will be produced and the designer should use the lowest composite in selecting an allowable load.

D-2. Example Computations. Included are examples for soil profiles consisting of soft clay, sand, silt, and alternating layers. To reduce the number of computations, a timber pile at a single tip elevation is shown. These computations must be made at each change in soil property and at a close enough interval to describe the curve mentioned in Paragraph D-1. In the examples, each of the various tip elevations used a timber pile having a 12-inch butt diameter and a 7-inch tip diameter, with the butt driven flush to the ground surface.

    a. Uniform Clay Profile. This example was developed for a bottom of slab elevation of +10.0 feet.* Using the shear strengths from the laboratory testing, two shear strength trends are developed. The example trend is shown below:

Average Laboratory Test Results

Elevation ft	φ°	Q Case γ (pcf)	c (psf)	φ°	S Case γ (pcf)	c (psf)
10.0 to 0.0	0	110	400	23	110	0
0.0 to -15.0	0	48*	600	23	48*	0
-15.0 to -20.0	0	42*	650	23	42*	0
-20.0 to -30.0	0	40*	800	23	40*	0
-30.0 to -60.0	0	38*	900	23	38*	0

    *The unit weight below the water table at elevation 0.0 (NGVD) is the
    submerged weight.

    (1) Example computation for soft clay in the "Q" case for a single tip elevation of -30.0 using a timber pile having butt diameter of 12 inches and tip diameter of 7 inches. This will be the computation for a single point on a pile capacity vs. tip elevation curve. It is based upon the data above which therefore means that the pile length is 40 ft.

    (a) Compute pile skin friction "Q" case. Computations will be by layer due to property variation computed as follows:

---

*All elevations (el) cited herein are in feet referred to National Geodetic
 Vertical Datum (NGVD) of 1929.

$$Q_s = \sum_{i=1 \text{ to } N} f_{s_i} A_{s_i}$$

To compute the average diameter of the tapered timber pile the following equation is used.

$$d_{i_A} = d_t + \frac{L(d_B - d_t)}{B}$$

where:

d$_{i_A}$ = diameter at the midpoint of the layer being computed

d$_t$ = diameter of pile tip

d$_B$ = diameter of pile butt

L = length from pile tip to midpoint of layer

L$_t$ = total length of pile

Layer from elevation +10.0 to elevation 0.0

Average pile diameter = $7 + \frac{35(5)}{40}$ = 11.375 inches

ΔQ_s = π · Avg. Diameter · Length · Cohesion · Adhesion reduction factor

obtained from Figure 4-3

Increment of skin friction

ΔQ_s = π $\frac{11.375}{12}$ (10)(400)(1) = 11,911 pounds

Layer from elevation 0.0 to elevation -15.0

Average pile diameter

$d = 7 + \frac{22.5(5)}{40}$ = 9.813 inches

α from Figure 4-3 is 0.95

Increment of skin friction

$$\Delta Q_s = \pi \frac{9.813}{12} (15')(600)(0.95) = 21,965 \text{ pounds}$$

Layer from elevation -15.0 to elevation -20.0

Average pile diameter

$$d = 7 + \frac{12.5(5)}{40} = 8.563 \text{ inches}$$

Increment of skin friction

$$\Delta Q_s = \pi \frac{8.563}{12} (5)(650)(1.9) = 6,556 \text{ pounds}$$

Layer from elevation -20.0 to elevation -30.0

Average pile diameter

$$d = 7 + \frac{5(5)}{40} = 7.625 \text{ inches}$$

Increment of skin friction

$$\Delta Q_s = \pi \frac{7.625}{12} (10)(800)(.8) = 12,712 \text{ pounds}$$

$$Q_s = \sum_{1-4} \Delta Q_s = 54,295 \text{ lb} = 27.15 \text{ tons}$$

$$Q_{S_{Allow}} = \frac{Q_S}{FS} = \frac{27.15 \text{ tons}}{2.0} = 13.57 \text{ tons}$$

(b) Compute end bearing "Q" case with pile tip at elevation -30.0 using the equations:

$$q = 9C$$

$$Q_T = A_T q$$

where:

C = shear strength

A_T = tip area

$$Q_T = 9.0(800) \frac{\pi(7/12)^2}{4} = 1,924.23 \text{ lb} = 0.962 \text{ tons}$$

$$Q_{T_{Allow}} = \frac{Q_T}{FS} = \frac{0.962}{2.0} = 0.48 \text{ tons}$$

The allowable "Q" case Compression soil/pile load with the pile tip at elevation -30.0 using a safety factor of 2.0 in the "Q" case is computed as follows:

$$Q_{A_{-30}} = Q_{S_A} + Q_{T_A} = 14 \text{ tons}$$

The allowable "Q" case tension soil/pile load with the pile tip at elevation -30.0 using a safety factor of 2.0 in the "Q" case is computed as follows:

$$Q_{A_{-30}} = Q_{S_A} \times k_T = 13.57 \text{ tons} \times 1.0 = 14 \text{ tons}$$

(2) Example computation for the "S" case capacity with all factors as in (1). This will compute a single point upon the "S" case portion of the curve pile capacity vs. tip elevation. The effective overburden pressure at any point has to be computed at any point in the soil profile, this is presented herein on page D-21, using the soil properties presented in the table above.

(a) Compute Skin Friction "S" case. Computations will be by the layer due to property variations as follows:

$$Q_s = \sum_{i=1 \text{ to } N} f_{s_i} A_{s_i}$$

where:

$f_s = k s_u$

$s_u = \gamma' D \tan \phi + C$

Layer from elevation +10.0 to elevation 0.0

Average pile diameter d = 11.375 inches (see previous computation)

Average strength (Note a reduction factor to the angle-of-internal friction ϕ is normally considered necessary to obtain δ). See Table 4-2 for values of δ to use. A reduction factor value of 1.0 is used herein. The engineer has to use judgement and load test results to find the range of δ for the area in which the work is being performed.

$$s_u = \frac{1,100 \tan 23°}{2} + 0 = 233.46 \text{ psf}$$

$s_u = 233.46$, $f_s = k\sigma' v = 233.46$ psf

Increment of skin friction

$$\Delta Q_s = \pi \frac{11.375}{12} (10)(233.46) = 6,952.4 \text{ pounds}$$

Layer from elevation 0.0 to elevation -15.0

Average pile diameter d = 9.813 inches

Average strength

$$s_u = \frac{1,100 + 1,820}{2} \tan 23° = 619.73$$

$f_s = k\sigma' v = 619.73$

D-7

Increment of skin friction

$$\Delta Q_s = \pi \ \frac{9.813}{12} \ (15)(619.73) = 23,880.43 \ \text{pounds}$$

Layer from elevation -15.0 to elevation -20.0

Average pile diameter   d = 8.563 inches

Average strength

$$s_u = \frac{1,820 + 2,030}{2} \ \tan 23° = 817.11 \ \text{psf}$$

$f_s = 817.11$ psf

Increment of skin friction

$$\Delta Q_s = \pi \ \frac{8.563}{12} \ (5)(817.11) = 9,158.94 \ \text{pounds}$$

Layer from elevation -20.0 to elevation -30.0

Average pile diameter   d = 7.625 inches

Average strength

$$s_u = \frac{2,030 + 2,430}{2} \ \tan 23° = 946.58 \ \text{psf}$$

$f_s = 946.58$ psf

Increment of skin friction

$$\Delta Q_s = \pi \ \frac{7.625}{12} \ (10)(946.58) = 18,895.82 \ \text{pounds}$$

$$Q_s = \sum_{i=1 \text{ to } N} \Delta Q_s = 58,887.6 \text{ lb} = 29.44 \text{ tons}$$

$$Q_{s_{\text{Allowable}}} = \frac{Q_s}{FS} = 14.72 \text{ tons}$$

(b) Compute end bearing in the "S" case with the pile tip at elevation -30.0 using the equations:

$$q = \sigma'_v N_q$$

$$Q_T = A_T q$$

where:

N_q = Terzaghi's bearing factor for ϕ = 23°

N_q = 10 (from "Terzaghi and Peck," Figure 4-2)

Area of tip = $A_T = \pi \dfrac{7/12^2}{4} = 0.267$ sq ft

Q_T = (0.267)(2430)(10) = 6,488.1 lb = 3.24 tons

$$Q_{s_{\text{Allowable}}} = \frac{Q_s}{FS} = 1.62 \text{ tons}$$

The allowable "S" case compression soil/pile load with the pile tip at elevation -30.0 using a safety factor of 2.0 in the "S" case is computed as follows:

$$Q_{A_{-30}} = Q_{S_A} + Q_{T_A} = 16.34 \text{ tons}$$

The allowable "S" case tension soil/pile load at elevation -30.0 tip, using a safety factor of 2.0 in the "S" case is as follows:

$$Q_{A_{-30}} = Q_{S_A} \times k_T = 14.72 \times 0.7 = 10.3 \text{ tons}$$

(3) Negative skin friction is directly addressed in other areas of this text and should be accounted for in any actual computation. Negative skin friction may occur, especially in the long-term, drained shear strength case. The soils engineer will estimate this load and furnish it to the structural engineer to be included into the applied loads.

e. Uniform Medium Density Sand Profile. This example was also developed for a bottom slab/ground line elevation of +10.0 feet, using the shear strengths from laboratory testing on a sand classified SP-F the shear strength trends shown below were developed.

Average Laboratory Test Results

Elevation (ft)	ϕ (degrees)	γ' (pcf)	c (psf)
10.0 to 0.0	30	122	0
0.0 to -60.0	30	60*	0

*The unit weight below the water table at el 0.0 is submerged weight.

(1) Example computation for sand using a single-pile tip elevation of -30.0 and a timber pile having butt diameter of 12 inches and tip diameter of 7 inches. This is a single point computation in a series to form a curve of pile tip elevation vs. pile capacity. This computation requires a 40 ft pile from elevation +10.0 to elevation -30.0.

(a) Compute the overburden pressure at the pile $\sigma'_v = \gamma D$ as shown in the computer example on Page D-24. Applying the critical depth limiting value criterion of D/B = 15 , the critical depth is 15 feet, taking B to be the butt diameter. Therefore, σ'_v is zero at the surface, 1,220 psf at elevation 0.0, and 1,520 psf at limit depth.

The angle of internal friction ϕ is reduced in some cases by δ shown in Table 4-3. In this example δ is taken as 1.0, which is based upon engineering judgement, the pile material and pile load tests in this area.

(b) Compute pile skin friction. Computations will be by layer due to property variations, as follows:

$$Q_s = \sum_{i=1 \text{ to } N} f_{s_i} A_{s_i}$$

where:

$f_{s_i} = k s_u$

$s_u = \gamma' D \tan \phi + c$

Layer from elevation +10.0 to elevation 0.0

Average pile diameter $d = 7 + \dfrac{35(5)}{40} = 11.375$ inches

D-10

Average strength of sand s_u

$$s_u = \frac{0 + 1,220}{2} \tan 30° + 0 = 352.18 \text{ psf}$$

$f_s = k s_u = 352.18$ psf

Increment of skin friction

$$\Delta Q_s = \pi \frac{11.375}{12} 10(352.18) = 10,487.83 \text{ pounds}$$

Layer from elevation 0.0 to elevation -5.0

Average pile diameter $d = 7 + \frac{27.5(5)}{40} = 10.438$ inches

Average strength of sand s_u

$$s_u = \frac{1,220 + 1,520}{2} \tan 30° = 790.97 \text{ psf}$$

$f_s = k s_u = 790.97$ psf

Increment of skin friction

$$\Delta Q_s = \pi \frac{10.438}{12} 5(790.97) = 10,806.78 \text{ pounds}$$

Layer from elevation -5.0 to elevation -30.0

Average pile diameter $d = 7 + \frac{12.5(5)}{40} = 8.563$ inches

Average strength of sand s_u

$$s_u = 1,520 \tan 30° = 877.57$$

$$f_s = ks_u = 877.57$$

$$\Delta Q_s = \pi \frac{8.563}{12} (25)(877.57) = 49,108.04 \text{ pounds}$$

$$\mathbf{Q_s = \sum_{i=1 \text{ to } N} \Delta Q_{s_i} = 70,475.0 \text{ pounds}}$$

$$Q_s = \qquad \Delta Q_{s_i} = 70,475.0 \text{ pounds}$$
$$i=1 \text{ to } N$$

(c) Compute end bearing with the tip at elevation -30.0 using the following equations:

$$q = \sigma_v' N_q$$

$$Q_T = A_T q$$

where:

N_q = Terzaghi's bearing factor for $\phi = 30°$

$N_q = 18$

Compute:

$\sigma_v' = 1,520$ (limit value at D/B =15)

$$A_T = \pi \frac{7/12^2}{4} = 0.2672 \text{ sq ft}$$

$Q_T = 7,311.24$ pounds

The allowable compression soil/pile load with the pile tip at elevation -30.0 using a safety factor of 2.0 will be:

$$Q_{A_{-30}} = \frac{Q_s + Q_T}{FS} = \frac{70,475.00 + 7,311.24}{2(2,000 \text{ lb/ton})} = 19.45 \text{ tons}$$

The allowable tension soil/pile load with the pile tip at elevation -30.0 using a safety factor of 2.0 will be:

D-12

$$Q_{A_{-30}} = \frac{Q_s \times k_T}{FS} = \frac{70,475(0.7)}{2(2,000 \text{ lb/ton})} = 12.33 \text{ tons}$$

c. Uniform Silt Subgrade. This example is developed for a bottom of slab/ground line elevation of +10.0 feet, using the shear strengths developed from laboratory $\bar{R}$ testing on a silt (ML). The "Q" values are from the envelope to the total stress circles, where the "S" values are from the effective stress circles. The silt used in this example was "dirty" with some clay included. The shear strength trends are as shown:

Average Laboratory Test Results

Elevation, ft	"Q" Case			"S" Case		
	ϕ (deg)	γ (pcf)	c (psf)	ϕ (deg)	γ (pcf)	c (psf)
10.0 to 0.0	15	117	200	28	117	0
0.0 to -60.0	15	55*	200	28	55*	0

*The unit weight (?) below the watertable at elevation 0.0 ft is the submerged unit weight.

(1) Example computation for silt in the "Q" case using a pile tip at elevation -30.0 and a timber pile having a butt diameter of 12 inches and a tip diameter of 7 inches. This is a single point computation in a series to form a curve of pile tip elevation vs. pile capacity. This computation will use a 40 foot pile to extend from elevation +10.0 to elevation -30.0.

(a) Compute the overburden pressure at the pile as $\sigma'_v = \gamma D$ as shown in the computer example on page D-27/28. Applying the critical depth limiting value criterion of $D/B = 15$ we find the critical depth to be 15 feet. Therefore, σ'_v at the surface in zero, at elevation 0.0, it is 1,170 psf, and at the limit depth elevation of -5.0, it is 1,445 psf.

(b) The angle of internal friction ϕ is reduced by a factor given in Table 4-3 to obtain δ. In this example δ is taken as 1.0 for the "Q" case and 0.9 for the "S" case. The use of reduction factor to obtain δ depends upon engineering judgement, the pile material, and the results of pile load tests in the area.

(c) Compute skin friction "Q" case. Computations will be by layer due to property variations as follows:

$$Q_s = \sum_{i=1 \text{ to } N} f_{s_i} A_{s_i}$$

where:

$$f_{s_i} = k s_u$$

D-13

$$s_u = \gamma'D \tan \phi + c$$

Layer from elevation +10.0 to elevation 0.0

Average pile diameter $d = 7 + \dfrac{35(5)}{40} = 11.375$ inches

Average strength of silt s_u

$$s_u = \frac{0 + 1,170}{2} \tan 15° + 200 = 356.72 \text{ psf}$$

$$f_s = ks_u = 356.72 \text{ psf}$$

Increment of skin friction

$$\Delta Q_s = \pi \frac{11.375}{12} (10)(356.72) = 63,738.19 \text{ pounds}$$

Layer from elevation 0.0 to elevation -5.0

Average pile diameter $d = 7 + \dfrac{27.5(5)}{40} = 10.438$ inches

Average strength of silt s_u

$$s_u = \frac{1,170 + 1,445}{2} \tan 15° + 200 = 550.34 \text{ psf}$$

$$f_s = ks_u = 550.34 \text{ psf}$$

Increment of skin friction

$$\Delta Q_s = \pi \frac{10.483}{12} (5)(550.34) = 7,519.48 \text{ pounds}$$

Layer from elevation -5.0 to elevation -30.0

D-14

Average pile diameter $d = 7 + \dfrac{12.5(5)}{40} = 8.563$ inches

Average strength of silt s_u

$s_u = 1,445 \tan 15° + 200 = 587.19$ psf

$f_s = ks_u = 587.19$ psf

Increment of skin friction

$\Delta Q_s = \pi \dfrac{8.563}{12} (25)(587.19) = 32,908.97$ pounds

$Q_s = \displaystyle\sum_{i=1 \text{ to } N} \Delta Q_{s_i} = 104,166.64 \text{ pounds}$

(d) Compute end bearing in the "Q" case with the pile tip at elevation -30.0 using the following equations:

$q = \sigma'_v N_q$

$Q_T = A_T q$

where:

N_q = Terzaghi's bearing factor at $\phi = 15°$

$N_q = 4$

compute:

$\sigma'_v = 1,445$ psf (limit value at D/B =15)

$A_T = \pi \dfrac{7/12^2}{4} = 0.2672$ sq ft

$Q_T = 0.2672(1,445)(4) = 1,544.42$ pounds

The allowable "Q" case compression soil/pile load with the pile tip at elevation -30.0 using a safety factor of 2.0 will be:

$$Q_{A_{-30}} = \frac{Q_S + Q_T}{FS} = 26.43 \text{ tons}$$

The allowable "Q" case tension soil/pile load with the pile tip at elevation -30.0 using a safety factor of 2.0 will be:

$$Q_{A_{-30}} = \frac{Q_S \times K_T}{FS} = 18.33 \text{ tons}$$

(2) Example computation for silt in the "S" case using a single pile tip at elevation -30.0 and a timber pile having a butt diameter of 12 inches and a tip diameter of 7 inches. This is a single point computation in a series to form a curve of pile tip elevation vs. pile capacity. This computation will use a 40-foot pile to extend from elevation +10.0 to elevation -30.0.

(a) The overburden pressure and limit values from the "Q" case example are valid; i.e., elevation +10.0, $\sigma'_v = 0.0$; elevation 0.0, $\sigma'_v = 1,170$ psf; elevation -5.0, $\sigma'_v = 1,445$ psf and $D/B_{critical} = 15$ feet.

(b) Compute skin friction "S" case. Computations will be by layer due to property variations as follows:

$$Q_s = \sum_{i=1 \text{ to } N} f_{s_i} A_{s_i}$$

where:

$$f_{s_i} = ks_u$$

$$s_u = \gamma'D \tan \phi + c$$

Layer from elevation +10.0 to elevation 0.0

$$\text{Average pile diameter} \quad d = 7 + \frac{35(5)}{40} = 11.375 \text{ inches}$$

Average strength of silt s_u

$$s_u = \frac{0 + 1170}{2} \tan (28° \times .9) + 0 = 275.28 \text{ psf}$$

$f_s = ks_u = 275.28$ psf

Increment of skin friction

$$\Delta Q_s = \pi \frac{11.375}{12} (10)(275.28) = 8,197.75 \text{ pounds}$$

Layer from elevation 0.0 to elevation -5.0

Average pile diameter $d = 7 + \dfrac{27.5(5)}{40} = 10.438$ inches

Average strength of silt s_u

$$s_u = \frac{1,170 + 1,445}{2} \tan (28° \times 0.9) = 615.26 \text{ psf}$$

$f_s = ks_u = 615.26$ psf

Increment of skin friction

$$\Delta Q_s = \pi \frac{10.438}{12} (5)(615.26) = 8,406.49 \text{ pounds}$$

Layer from elevation -5.0 to elevation -30.0

Average pile diameter $d = 7 + \dfrac{12.5(5)}{40} = 8.563$ inches

Average strength of silt s_u

$$s_u = (1,445) \tan (28 \times 0.9) = 679.97 \text{ psf}$$

D-17

$f_s = ks_u = 679.97$ psf

Increment of skin friction

$$\Delta Q_s = \pi \frac{8.563}{12} (25)(679.97) = 38,108.71 \text{ pounds}$$

$$Q_s = \sum_{i=1 \text{ to N}} \Delta Q_s = 54,712.95 \text{ pounds}$$

(c)   Compute end bearing in the "S" case with the tip of at elevation -30.0 using the following equations:

$$q = \sigma'_v N_q$$

$$Q_T = A_T q$$

where:

N_q = Terzaghi's bearing factor at $\phi = 28°$

$N_q = 15$

Compute:

$\sigma'_v = 1,445$ psf (limit value of D/B = 15)

$$A_T = \pi \frac{7/12^2}{4} = 0.2672 \text{ sq ft}$$

$$Q_T = 0.2672(1,445)(15) = 5,791.56 \text{ pounds}$$

The allowable "S" case compression soil/pile load with the pile tip at elevation -30.0, using a safety factor of 2.0, will be:

$$Q_{A_{-30}} = \frac{Q_s + Q_T}{FS} = 15.13 \text{ tons}$$

The allowable "S" case tension soil/pile load with the pile tip at elevation -30.0, using a safety factor of 2.0, will be:

$$Q_A{}_{-30} = \frac{Q_s \times K_T}{FS} = \frac{54,712.95(.7)}{2,(2,000 \text{ lb/ton})} = 9.57 \text{ tons}$$

f. Layered Clay, Silt and Sand Subgrade. This example is developed for a bottom of slab/ground line elevation of +10.0 feet, using the strengths developed from laboratory testing on the various soil types as follows:

Average Laboratory Test Results

Elevation (ft)	"Q" case			"S" case			Soil Type
	ϕ (deg)	γ (pcf)	c (psf)	ϕ (deg)	γ (pcf)	c (psf)	
+10.0 to 0.0	0	110	400	23	110	0	CH
0.0 to -12.0	15	55*	200	28	55*	0	ML
-12.0 to -20.0	0	38*	600	23	38*	0	CH
-20.0 to -60.0	30	60*	0	30	60*	0	SP

*The unit weight (γ) below the water table at elevation 0.0 feet is the submerged weight.

(1) Example computation for a multiplayered soil in the "Q" case using a single pile tip at elevation -30.0 and a timber pile having a butt diameter of 12 inches and a tip diameter of 7 inches. This is a single-point computation in a series to form a curve of pile tip elevation vs. pile capacity. This computation will use a 40-foot pile to extend from elevation +10.0 to elevation -30.0.

(a) Compute the overburden pressure at the pile as $\sigma'_v = \gamma D$ as shown on the computer example on Page D-31. It should be noted that the critical depth limit was applied from the upper surface of the granular layer and not from the ground surface.

(b) The angle of internal friction ϕ is reduced by a factor given in Table 4-3 to obtain δ. In this example, δ is taken as 1.0 for both the "Q" and "S" case. The use of reduction factor to obtain δ depends upon engineering judgement, the pile material and results of pile load test in the area.

(c) Compute Skin Friction "Q" Case. Computations will be by layer due to material variations as follows:

$$Q_s = \sum_{i=1 \text{ to } N} f_{s_i} A_{s_i}$$

where:

$$f_{s_i} = ks_u$$

$$s_u = \gamma'D \tan \phi + c$$

Layer from elevation +10.0 to 0.0 (clay)

Average pile diameter $\quad d = 7 + \dfrac{35(5)}{40} = 11.375$

Average shear strength $\quad c = 400 \text{ psf} = f_s$

Increment of skin friction

$$\Delta Q_s = \pi \, \dfrac{11.375}{12} \, (10)(400) = 11,911.9 \text{ pounds}$$

Layer from elevation 0.0 to elevation -12.0 (silt)

Average pile diameter $\quad d = 7 + \dfrac{24(5)}{40} = 10.0 \text{ inches}$

σ'_v top of stratum = 10' × 110 pcf = 1,100 psf

σ'_v bottom of stratum = 10' × 110 pcf + 12' × 55 pcf = 1,760 psf

Average strength in silt $\quad s_u$

$$s_u = \dfrac{1,100 + 1,760}{2} \tan 15° + 200 = 583.17 \text{ psf}$$

$$f_s = ks_u = 583.17 \text{ psf}$$

Increment of skin friction

$$\Delta Q_s = \pi \, \dfrac{10.0}{12} \, (12)(583.17) = 18,320.87 \text{ pounds}$$

D-20

Layer from elevation -12.0 to elevation -20.0 (clay)

Average pile diameter $\quad d = 7 + \dfrac{23(5)}{40} = 9.875$ inches

Average shear strength $\quad C = 600$ psf $= f_s$

Increment of skin friction

$\Delta Q_s = \pi \dfrac{9.875}{12} (8)(600)(0.95) = 11{,}788.85$ pounds

Layer from elevation -20.0 to elevation -30.0 (sand)

Average pile diameter $\quad d = 7 + \dfrac{5(5)}{40} = 7.625$ inches

σ'_v top of stratum $= 10' \times 110$ pcf $+ 12' \times 55$ pcf $+ 8' \times 38$ pcf

$= 2{,}064$ psf

σ'_v bottom of stratum $= 10' \times 110$ pcf $+ 12' \times 55$ pcf $+ 8 \times 38$ pcf $+ 10'$

$\times 60$ pcf $= 2{,}664$ psf

Average strength of sand $\quad s_u$

$s_u = \dfrac{2{,}064 + 2{,}664}{2} \tan 30° = 1{,}364.86$ psf

$f_s = k s_u = 1{,}364.86$ psf

Increment of skin friction

$\Delta Q_s = \pi \dfrac{7.625}{12} (10)(1{,}364.86) = 27{,}245.68$ pounds

$$Q_s = \sum_{i=1 \text{ to } N} \Delta Q_s = 69,887.77 \text{ pounds}$$

(d) Compute end bearing in the "Q" case with the pile tip at elevation -30.0 using the following equations:

$$q = \sigma_v' N_q$$

$$Q_T = A_T q$$

where:

N_q = Terzaghi's bearing capacity factor N_q = 18 at ϕ = 30

σ_v' = 2,664 psf (limit value is greater than 10 feet)
(Refer to paragraph C-2.f(1)(2))

$$A_T = \pi \frac{7/12^2}{4} = 0.2672 \text{ sq ft}$$

Q_T = 0.2672(2664)(18) = 12,812.77 pounds

The allowable "Q" case compression soil/pile load with the pile tip at elevation -30.0 and a safety factor of 2.0 will be:

$$Q_{A_{-30}} = \frac{Q_s + Q_T}{FS} = 20.53 \text{ tons}$$

The allowable "Q" case tension soil/pile load with the pile tip at elevation -30.0 and a safety factor will be:

$$Q_{A_{-30}} = \frac{Q_s \times K_T}{FS} = 20.13 \text{ tons}$$

(2) Example computation for a multilayered system in the "S" case using a single pile tip elevation -30.0 and a timber pile having a butt diameter of 12 inches and a tip diameter of 7 inches. This is a single point computation in a series to form a curve of pile tip elevation vs. pile capacity. This computation will use a 40 foot pile to extend from elevation +10.0 to -30.0.

(a) Compute the overburden pressure as in Paragraph 1.a. The angle-of-internal friction (ϕ) reduction factor is reduced by a factor given in Table 4-3 to obtain δ is taken as 1.0 as discussed in Paragraph 1.b.

D-22

(b) Compute Skin Friction "S" case. Computations will be by layer due to material variations as follows:

$$Q_s = \sum_{i=1 \text{ to } N} f_{s_i} A_{s_i}$$

where:

$$f_{s_i} = ks_u$$

$$s_u = \gamma'D \tan \phi + c$$

Layer from elevation +10.0 to elevation 0.0 (clay)

Average pile diameter $d = 7 + \dfrac{35(5)}{40} = 11.375$ inches

σ'_v top of stratum = 0

σ'_v bottom of stratum = 10' × 110 pcf = 1,100 psf

Average strength in clay s_u

$$s_u = \frac{0 + 1,100}{2} \tan 23° + 0 = 233.46 \text{ psf}$$

$$f_s = ks_u = 233.46 \text{ psf}$$

Increment of skin friction

$$\Delta Q_s = \pi \frac{11.375}{12}(10)(233.46) = 6,952.38 \text{ pounds}$$

Layer from elevation 0.0 to elevation -12.0 (silt)

Average pile diameter $d = 7 + \dfrac{24(5)}{40} = 10.0$ inches

σ'_v top of stratum = 10' × 110 pcf = 1,100 psf

σ'_v bottom of stratum = 10' × 110 pcf + 12' × 55 pcf = 1,760 psf

Average strength in silt s_u

$$s_u = \frac{1100 + 1,760}{2}\tan 28° = 760.34 \text{ psf}$$

$f_s = ks_u = 760.34$ psf

Increment of skin friction

$$\Delta Q_s = \pi \frac{10.0}{12} (12)(760.34) = 23,886.84 \text{ pounds}$$

Layer from elevation -12.0 to elevation -20.0 (clay)

Average pile diameter $d = 7 + \frac{23(5)}{40} = 9.875$ inches

σ'_v top of stratum = 10' × 110 pcf + 12' × 55 pcf = 1,760 psf

σ'_v bottom of stratum = 10' × 110 pcf + 12' × 55 pcf + 8' × 38 pcf

$$= 2,064 \text{ psf}$$

Average strength in clay s_u

$$s_u = \frac{1,760 + 2,064}{2}\tan 23° = 811.6 \text{ psf}$$

$f_s = ks_u = 811.6$ psf

Increment of skin friction

$$\Delta Q_s = \pi \frac{9.875}{12} (8)(811.6) = 16,785.67 \text{ pounds}$$

Layer from elevation -20.0 to elevation -30.0 (sand)

Average pile diameter $d = 7 + \dfrac{5(5)}{40} = 7.625$ inches

σ'_v top of stratum = 10' × 110 pcf + 12' × 55 pcf + 8' × 38 pcf

$$= 2,064 \text{ psf}$$

σ'_v bottom of stratum = 10' × 110 pcf + 12' × 55 pcf + 8' × 38 pcf + 10' × 60 pcf = 2,664 psf

Average strength in sand s_u

$$s_u = \frac{2,064 + 2,664}{2} \tan 30° = 1,364.86 \text{ psf}$$

$f_s = ks_u = 1,364.86$ psf

Increment of skin friction

$$\Delta Q_s = \pi \frac{7.625}{12} (10)(1,364.86) = 27,245.68 \text{ pounds}$$

$$\mathbf{Q_s = \sum_{i=1 \text{ to } N} \Delta Q_s = 74,870.57 \text{ pounds}}$$

(c) Compute end bearing in "S" case with the pile tip at elevation -30.0 using the following equations:

$$q = \sigma'_v N_q$$

$$Q_T = A_T q$$

where:

N_q = Terzaghi's Bearing Capacity Chart at $\phi = 30°$

$N_q = 18$

$\sigma'_v = 2,664$ psf (limit value greater than 10 feet)

$$A_T = \pi \frac{7/12^2}{4} = 0.2672 \text{ sq ft}$$

$$Q_T = 0.2672(2,664)(18) = 12,812.77 \text{ pounds}$$

The allowable "S" case compression soil/pile load with the pile tip at elevation -30.0 and a safety factor of 2.0 will be:

$$Q_{A_{-30}} = \frac{Q_s + Q_T}{FS} = 21.92 \text{ tons}$$

The allowable "S" case tension soil/pile load with the pile tip at elevation -30.0 and a safety factor of 2.0 will be:

$$Q_{A_{-30}} = \frac{Q_s \times K_T}{FS} = 13.10 \text{ tons}$$

e. Computer programs are currently available to compute pile capacity, examples similar to those computed above are shown in Tables C1 through C4.

Table D-1

Computer Output for Allowable Design Loads
for Uniform Soft Clay Subgrade

---

* PILE CAPACITY COMPUTATIONS *

CASE I
TIMBER PILE
CLAY SUBGRADE

****
CLASS B TIMBER PILE
****

PILE BUTT DIA. IS AT THE GROUND SURFACE FOR ALL TIP PENETRATIONS
PILE LENGTH USED IS FROM GROUND SURFACE TO TIP
TIMBER PILE DIM.: 12.0 IN.BUTT DIA.,   7.0 IN.TIP DIA.
TOP SURFACE ELEV.:,   10.0 FT.

**** Q-CASE ****
----

* STRATUM NUMBER *	SOIL/SOIL FRIC.ANG DEG.	WEIGHT DENSITY LB/CU FT	TOP STRA COHESION LB/SQ FT	BOT STRA COHESION LB/SQ FT	COEF LAT EAR.PRES (KC)
1	0.00	110.00	400.00	400.00	1.00
2	0.00	48.00	400.00	400.00	1.00
3	0.00	48.00	600.00	600.00	1.00
4	0.00	42.00	650.00	650.00	1.00
5	0.00	40.00	800.00	800.00	1.00
6	0.00	38.00	900.00	900.00	1.00

* STRATUM NUMBER *	COEF LAT EAR.PRES (KT)	TERZAGHI BEAR.CAP. (NC)	TERZAGHI BEAR.CAP. (NQ)	BOTTOM ELEVATION FEET	SOIL/PILE FRIC.ANG. DEG.	SOIL TYPE
1	0.70	9.00	1.00	0.00		CH
2	0.70	9.00	1.00	-5.00		CH
3	0.70	9.00	1.00	-15.00		CH
4	0.70	9.00	1.00	-20.00		CH
5	0.70	9.00	1.00	-30.00		CH
6	0.70	9.00	1.00	-60.00		CH

(Continued)

D-27

Table D-1 (Continued)

STRAT *NUM *	EL.TIP (FT)	COH/ADH. RESISTAN TONS	FRICTION COMPRESS TONS	RESISTAN TENSION TONS	END BEARING TONS	PILE CAPAC IN COMPRS. TONS	PILE CAPAC IN TENSION TONS
1	5.000	2.487	0.000	0.000	0.555	3.042	2.487
1	0.000	4.974	0.000	0.000	0.628	5.602	4.974
2	0.000				0.628	5.602	4.974
2	-2.500	6.218	0.000	0.000	0.644	6.862	6.218
2	-5.000	7.461	0.000	0.000	0.660	8.121	7.461
3	-5.000				0.901	8.362	7.461
3	-10.000	10.946	0.000	0.000	0.901	11.847	10.946
3	-15.000	14.530	0.000	0.000	0.901	15.431	14.530
4	-15.000				0.961	15.491	14.530
4	-17.500	16.460	0.000	0.000	0.961	17.421	16.460
4	-20.000	18.405	0.000	0.000	0.961	19.366	18.405
5	-20.000				1.141	19.546	18.405
5	-25.000	23.050	0.000	0.000	1.141	24.191	23.050
5	-30.000	27.777	0.000	0.000	1.141	28.918	27.777
6	-30.000				1.261	29.039	27.777
6	-45.000	43.737	0.000	0.000	1.261	44.998	43.737
6	-60.000	60.051	0.000	0.000	1.261	61.313	60.051

**** MODIFIED S-CASE ****
----

* STRATUM * NUMBER *	SOIL/SOIL FRIC.ANG DEG.	WEIGHT DENSITY LB/CU FT	TOP STRA COHESION LB/SQ FT	BOT STRA COHESION LB/SQ FT	COEF LAT EAR.PRES (KC)
1	23.00	110.00	0.00	0.00	1.00
2	23.00	48.00	0.00	0.00	1.00
3	23.00	48.00	0.00	0.00	1.00
4	23.00	42.00	0.00	0.00	1.00
5	23.00	40.00	0.00	0.00	1.00
6	23.00	38.00	0.00	0.00	1.00

* STRATUM * NUMBER *	COEF LAT EAR.PRES (KT)	TERZAGHI BEAR.CAP. (NC)	TERZAGHI BEAR.CAP. (NQ)	BOTTOM ELEVATION FEET	SOIL/PILE FRIC.ANG. DEG.	SOIL TYPE
1	0.70	0.00	10.00	0.00		CH
2	0.70	0.00	10.00	-5.00		CH
3	0.70	0.00	10.00	-15.00		CH
4	0.70	0.00	10.00	-20.00		CH
5	0.70	0.00	10.00	-30.00		CH
6	0.70	0.00	10.00	-60.00		CH

(Continued)

Table D-1 (Concluded)

STRAT *NUM *	EL.TIP (FT)	COH/ADH. RESISTAN TONS	FRICTION COMPRESS TONS	RESISTAN TENSION TONS	END BEARING TONS	PILE CAPAC IN COMPRS. TONS	PILE CAPAC IN TENSION TONS
1	5.000	0.000	0.726	0.508	0.735	1.461	0.508
1	0.000	0.000	2.903	2.032	1.470	4.373	2.032
2	0.000				1.470	4.373	2.032
2	-2.500	0.000	4.265	2.985	1.630	5.895	2.985
2	-5.000	0.000	5.813	4.069	1.791	7.603	4.069
3	-5.000				1.791	7.603	4.069
3	-10.000	0.000	9.132	6.392	1.791	10.922	6.392
3	-15.000	0.000	12.538	8.777	1.791	14.329	8.777
4	-15.000				1.791	14.329	8.777
4	-17.500	0.000	14.259	9.981	1.791	16.049	9.981
4	-20.000	0.000	15.988	11.191	1.791	17.778	11.191
5	-20.000				1.791	17.778	11.191
5	-25.000	0.000	19.462	13.623	1.791	21.253	13.623
5	-30.000	0.000	22.952	16.066	1.791	24.743	16.066
6	-30.000				1.791	24.743	16.066
6	-45.000	0.000	33.473	23.431	1.791	35.263	23.431
6	-60.000	0.000	44.032	30.822	1.791	45.822	30.822

* RUN COMPLETED *   1 TON = 2000 LBS.

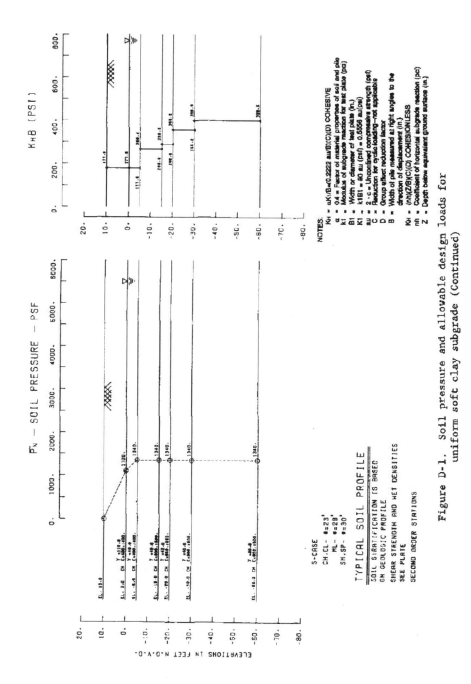

Figure D-1. Soil pressure and allowable design loads for
uniform soft clay subgrade (Continued)

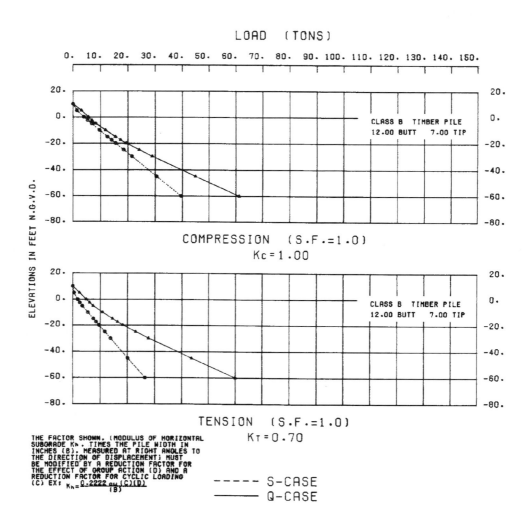

Figure D-1.   (Concluded)

Table D-2

Computer Output for Allowable Design Loads for
Uniform Medium Density Sand Subgrade, Run 1

---

* PILE CAPACITY COMPUTATIONS *

CASE II
TIMBER PILE
SAND SUBGRADE

****
CLASS B TIMBER PILE
****

PILE BUTT DIA. IS AT THE GROUND SURFACE FOR ALL TIP PENETRATIONS
PILE LENGTH USED IS FROM GROUND SURFACE TO TIP
TIMBER PILE DIM.: 12.0 IN.BUTT DIA., 7.0 IN.TIP DIA.
TOP SURFACE ELEV.:, 10.0 FT.

**** Q-CASE ****
----

* STRATUM * NUMBER *	SOIL/SOIL FRIC.ANG DEG.	WEIGHT DENSITY LB/CU FT	TOP STRA COHESION LB/SQ FT	BOT STRA COHESION LB/SQ FT	COEF LAT EAR.PRES (KC)
1	30.00	122.00	0.00	0.00	1.00
2	30.00	60.00	0.00	0.00	1.00
3	30.00	60.00	0.00	0.00	1.00
4	30.00	60.00	0.00	0.00	1.00

* STRATUM * NUMBER *	COEF LAT EAR.PRES (KT)	TERZAGHI BEAR.CAP. (NC)	TERZAGHI BEAR.CAP. (NQ)	BOTTOM ELEVATION FEET	SOIL/PILE FRIC.ANG. DEG.	SOIL TYPE
1	0.70	9.00	18.00	0.00		SP
2	0.70	9.00	18.00	-5.00		SP
3	0.70	9.00	18.00	-30.00		SP
4	0.70	9.00	18.00	-60.00		SP

(Continued)

STRAT CAPAC *NUM TENSION * TONS	EL.TIP (FT)	COH/ADH. RESISTAN	FRICTION COMPRESS TONS	RESISTAN TENSION TONS	END BEARING TONS	PILE CAPAC IN COMPRS. TONS	PILE IN TONS
1	5.000	0.000	1.095	0.766	1.467	2.562	0.766
1	0.000	0.000	4.380	3.066	2.934	7.314	3.066
2	0.000				2.934	7.314	3.066
2	-2.500	0.000	6.445	4.512	3.295	9.740	4.512
2	-5.000	0.000	8.819	6.173	3.656	12.475	6.173
3	-5.000				3.656	12.475	6.173
3	-17.500	0.000	21.832	15.282	3.656	25.488	15.282
3	-30.000	0.000	35.238	24.666	3.656	38.894	24.666
4	-30.000				3.656	38.894	24.666
4	-45.000	0.000	51.466	36.026	3.656	55.122	36.026
4	-60.000	0.000	67.754	47.428	3.656	71.410	47.428

**** MODIFIED S-CASE ****
----

* STRATUM * NUMBER *	SOIL/SOIL FRIC.ANG DEG.	WEIGHT DENSITY LB/CU FT	TOP STRA COHESION LB/SQ FT	BOT STRA COHESION LB/SQ FT	COEF LAT EAR.PRES (KC)
1	30.00	122.00	0.00	0.00	1.00
2	30.00	60.00	0.00	0.00	1.00
3	30.00	60.00	0.00	0.00	1.00
4	30.00	60.00	0.00	0.00	1.00

* STRATUM * NUMBER *	COEF LAT EAR.PRES (KT)	TERZAGHI BEAR.CAP. (NC)	TERZAGHI BEAR.CAP. (NQ)	BOTTOM ELEVATION FEET	SOIL/PILE FRIC.ANG. DEG.	SOIL TYPE
1	0.70	0.00	10.00	0.00		SP
2	0.70	0.00	10.00	-5.00		SP
3	0.70	0.00	10.00	-30.00		SP
4	0.70	0.00	10.00	-60.00		SP

(Continued)

Table D-2 (Concluded)

STRAT CAPAC *NUM TENSION * TONS	EL.TIP (FT)	COH/ADH. RESISTAN	FRICTION COMPRESS TONS	RESISTAN TENSION TONS	END BEARING TONS	PILE CAPAC IN COMPRS. TONS	PILE IN TONS
1	5.000	0.000	1.095	0.766	0.815	1.910	0.766
1	0.000	0.000	4.380	3.066	1.630	6.010	3.066
2	0.000				1.630	6.010	3.066
2	-2.500	0.000	6.445	4.512	1.831	8.276	4.512
2	-5.000	0.000	8.819	6.173	2.031	10.850	6.173
3	-5.000				2.031	10.850	6.173
3	-17.500	0.000	21.832	15.282	2.031	23.863	15.282
3	-30.000	0.000	35.238	24.666	2.031	37.269	24.666
4	-30.000				2.031	37.269	24.666
4	-45.000	0.000	51.466	36.026	2.031	53.497	36.026
4	-60.000	0.000	67.754	47.428	2.031	69.786	47.428

* RUN COMPLETED *   1 TON = 2000 LBS.

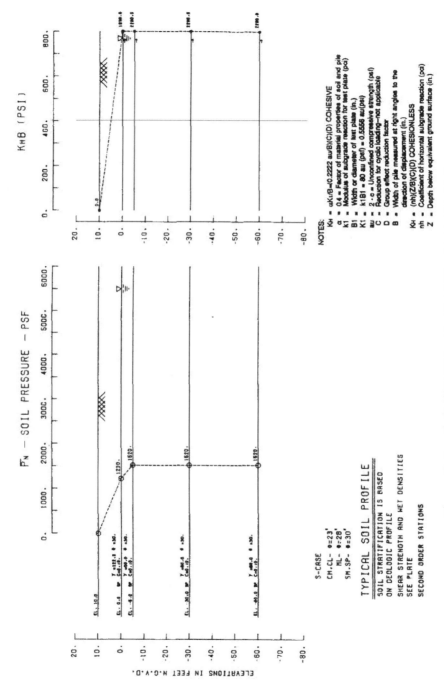

Figure D-2. Soil pressure and allowable design loads for uniform medium density sand subgrade (Continued)

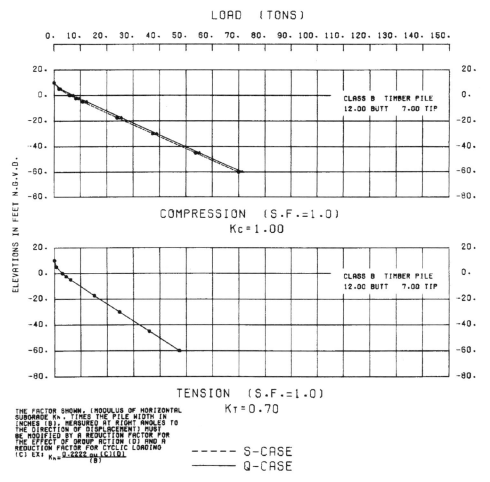

LOAD (TONS)

0. 10. 20. 30. 40. 50. 60. 70. 80. 90. 100. 110. 120. 130. 140. 150.

ELEVATIONS IN FEET N.G.V.D.

CLASS B TIMBER PILE
12.00 BUTT   7.00 TIP

COMPRESSION (S.F.=1.0)
Kc=1.00

CLASS B TIMBER PILE
12.00 BUTT   7.00 TIP

TENSION (S.F.=1.0)
KT=0.70

THE FACTOR SHOWN, (MODULUS OF HORIZONTAL
SUBGRADE Kh, TIMES THE PILE WIDTH IN
INCHES (B), MEASURED AT RIGHT ANGLES TO
THE DIRECTION OF DISPLACEMENT) MUST
BE MODIFIED BY A REDUCTION FACTOR FOR
THE EFFECT OF GROUP ACTION (D) AND A
REDUCTION FACTOR FOR CYCLIC LOADING
(C) EX: $K_h = \frac{0.2222}{(B)} ou (C)(D)$

- - - - - S-CASE
————— Q-CASE

Figure D-2. (Concluded)

Computer Output for Allowable Design Loads
for Uniform Silt Subgrade, Run 2

---

* PILE CAPACITY COMPUTATIONS *

CASE III
TIMBER PILE

****
CLASS B   TIMBER PILE
****

PILE BUTT DIA. IS AT THE GROUND SURFACE FOR ALL TIP PENETRATIONS
PILE LENGTH USED IS FROM GROUND SURFACE TO TIP
TIMBER PILE DIM.: 12.0 IN.BUTT DIA.,   7.0 IN.TIP DIA.
TOP SURFACE ELEV.:,   10.0 FT.

**** Q-CASE ****

* STRATUM * NUMBER *	SOIL/SOIL FRIC.ANG DEG.	WEIGHT DENSITY LB/CU FT	TOP STRA COHESION LB/SQ FT	BOT STRA COHESION LB/SQ FT	COEF LAT EAR.PRES (KC)
1	15.00	117.00	200.00	200.00	1.00
2	15.00	55.00	200.00	200.00	1.00
3	15.00	55.00	200.00	200.00	1.00
4	15.00	55.00	200.00	200.00	1.00

* STRATUM * NUMBER *	COEF LAT EAR.PRES (KT)	TERZAGHI BEAR.CAP. (NC)	TERZAGHI BEAR.CAP. (NQ)	BOTTOM ELEVATION FEET	SOIL/PILE FRIC.ANG. DEG.	SOIL TYPE
1	0.70	9.00	4.00	0.00		ML
2	0.70	9.00	4.00	-5.00		ML
3	0.70	9.00	4.00	-30.00		ML
4	0.70	9.00	4.00	-60.00		ML

STRAT *NUM *	EL.TIP (FT)	COH/ADH. RESISTAN TONS	FRICTION COMPRESS TONS	RESISTAN TENSION TONS	END BEARING TONS	PILE CAPAC IN COMPRS. TONS	PILE CAPAC IN TENSION TONS
1	5.000	1.244	0.487	0.341	0.553	2.284	1.585
1	0.000	2.487	1.949	1.364	0.866	5.302	3.852
2	0.000				0.866	5.302	3.852
2	-2.500	3.109	2.867	2.007	0.939	6.915	5.115
2	-5.000	3.731	3.916	2.742	1.013	8.660	6.472
3	-5.000				1.013	8.660	6.472
3	-17.500	6.840	9.661	6.763	1.013	17.513	13.602
3	-30.000	9.948	15.577	10.904	1.013	26.538	20.852
4	-30.000				1.013	26.538	20.852
4	-45.000	13.679	22.737	15.916	1.013	37.429	29.595
4	-60.000	17.410	29.924	20.947	1.013	48.347	38.357

(Continued)

Table D-3 (Concluded)

**** MODIFIED S-CASE ****
----

* STRATUM * NUMBER *	SOIL/SOIL FRIC.ANG DEG.	WEIGHT DENSITY LB/CU FT	TOP STRA COHESION LB/SQ FT	BOT STRA COHESION LB/SQ FT	COEF LAT EAR.PRES (KC)
1	28.00	117.00	0.00	0.00	1.00
2	28.00	55.00	0.00	0.00	1.00
3	28.00	55.00	0.00	0.00	1.00
4	28.00	55.00	0.00	0.00	1.00

* STRATUM * NUMBER *	COEF LAT EAR.PRES (KT)	TERZAGHI BEAR.CAP. (NC)	TERZAGHI BEAR.CAP. (NQ)	BOTTOM ELEVATION FEET	SOIL/PILE FRIC.ANG. DEG.	SOIL TYPE
1	0.70	0.00	10.00	0.00		ML
2	0.70	0.00	10.00	-5.00		ML
3	0.70	0.00	10.00	-30.00		ML
4	0.70	0.00	10.00	-60.00		ML

STRAT *NUM *	EL.TIP (FT)	COH/ADH. RESISTAN TONS	FRICTION COMPRESS TONS	RESISTAN TENSION TONS	END BEARING TONS	PILE CAPAC IN COMPRS. TONS	PILE CAPAC IN TENSION TONS
1	5.000	0.000	0.967	0.677	0.782	1.749	0.677
1	0.000	0.000	3.868	2.708	1.563	5.431	2.708
2	0.000				1.563	5.431	2.708
2	-2.500	0.000	5.688	3.982	1.747	7.435	3.982
2	-5.000	0.000	7.772	5.440	1.931	9.703	5.440
3	-5.000				1.931	9.703	5.440
3	-17.500	0.000	19.171	13.419	1.931	21.101	13.419
3	-30.000	0.000	30.910	21.637	1.931	32.840	21.637
4	-30.000				1.931	32.840	21.637
4	-45.000	0.000	45.119	31.583	1.931	47.050	31.583
4	-60.000	0.000	59.380	41.566	1.931	61.311	41.566

* RUN COMPLETED * 1 TON = 2000 LBS.

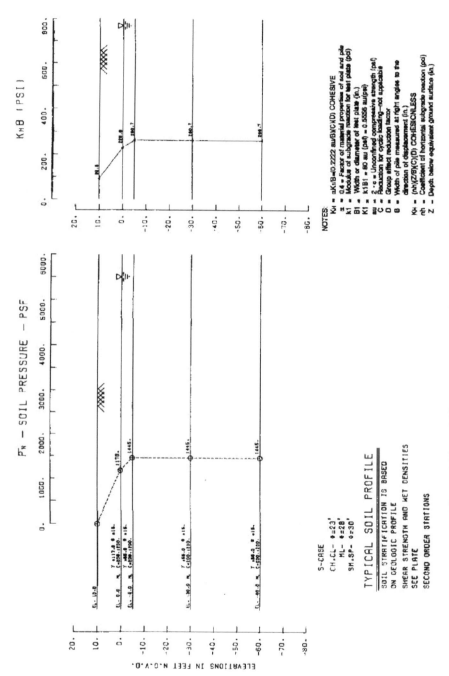

Figure D-3. Soil pressure and allowable design loads for
uniform silt subgrade (Continued)

LOAD (TONS)

0. 10. 20. 30. 40. 50. 60. 70. 80. 90. 100. 110. 120. 130. 140. 150.

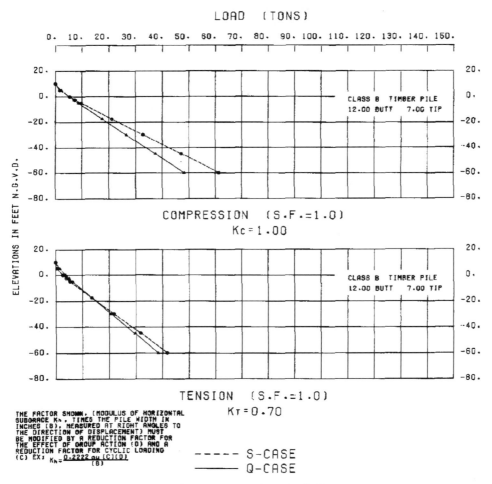

COMPRESSION (S.F.=1.0)
Kc=1.00

TENSION (S.F.=1.0)
KT=0.70

THE FACTOR SHOWN, (MODULUS OF HORIZONTAL
SUBGRADE Kₕ, TIMES THE PILE WIDTH IN
INCHES (B), MEASURED AT RIGHT ANGLES TO
THE DIRECTION OF DISPLACEMENT) MUST
BE MODIFIED BY A REDUCTION FACTOR FOR
THE EFFECT OF GROUP ACTION (G) AND A
REDUCTION FACTOR FOR CYCLIC LOADING
(C) EX: $K_h = \frac{0.2222 \text{ au } (C)(G)}{(B)}$

- - - - - S-CASE
───────── Q-CASE

Figure D-3. (Concluded)

Table D-4

Computer Output for Allowable Design Loads for
Layered Clay, Silt, and Sand Subgrade, Run 3

---

* PILE CAPACITY COMPUTATIONS *

CASE IV
TIMBER PILE
INTERBEDDED SOILS

****
CLASS B TIMBER PILE
****

PILE BUTT DIA. IS AT THE GROUND SURFACE FOR ALL TIP PENETRATIONS
PILE LENGTH USED IS FROM GROUND SURFACE TO TIP
TIMBER PILE DIM.: 12.0 IN.BUTT DIA., 7.0 IN.TIP DIA.
TOP SURFACE ELEV.:, 10.0 FT.

**** Q-CASE ****
----

* STRATUM NUMBER *	SOIL/SOIL FRIC.ANG DEG.	WEIGHT DENSITY LB/CU FT	TOP STRA COHESION LB/SQ FT	BOT STRA COHESION LB/SQ FT	COEF LAT EAR.PRES (KC)
1	0.00	110.00	400.00	400.00	1.00
2	15.00	55.00	200.00	200.00	1.00
3	15.00	55.00	200.00	200.00	1.00
4	0.00	38.00	600.00	600.00	1.00
5	30.00	60.00	0.00	0.00	1.00

* STRATUM NUMBER *	COEF LAT EAR.PRES (KT)	TERZAGHI BEAR.CAP. (NC)	TERZAGHI BEAR.CAP. (NQ)	BOTTOM ELEVATION FEET	SOIL/PILE FRIC.ANG. DEG.	SOIL TYPE
1	0.70	9.00	1.00	0.00		CH
2	0.70	9.00	4.00	-5.00		ML
3	0.70	9.00	4.00	-12.00		ML
4	0.70	9.00	1.00	-20.00		CH
5	0.70	9.00	18.00	-60.00		SP

(Continued)

D-41

Table D-4 (Continued)

STRAT CAPAC *NUM TENSION TONS	EL.TIP (FT)	COH/ADH. RESISTAN	FRICTION COMPRESS TONS	RESISTAN TENSION TONS	END BEARING TONS	PILE CAPAC IN COMPRS. TONS	PILE IN TONS
1	5.000	2.487	0.000	0.000	0.555	3.042	2.487
1	0.000	4.974	0.000	0.000	0.628	5.602	4.974
2	0.000				0.828	5.803	4.974
2	-2.500	5.727	0.769	0.538	0.902	7.397	6.265
2	-5.000	6.436	1.700	1.190	0.975	9.111	7.626
3	-5.000				0.975	9.111	7.626
3	-8.500	7.389	3.132	2.193	0.975	11.497	9.582
3	-12.000	8.316	4.619	3.234	0.975	13.911	11.549
4	-12.000				0.905	13.840	11.549
4	-16.000	10.903	4.933	3.453	0.905	16.741	14.356
4	-20.000	13.596	5.162	3.614	0.905	19.664	17.210
5	-20.000				3.307	22.066	17.210
5	-40.000	15.195	22.386	15.670	3.307	40.888	30.865
5	-60.000	15.880	41.050	28.735	3.307	60.237	44.615

**** MODIFIED S-CASE ****
----

* STRATUM * NUMBER *	SOIL/SOIL FRIC.ANG DEG.	WEIGHT DENSITY LB/CU FT	TOP STRA COHESION LB/SQ FT	BOT STRA COHESION LB/SQ FT	COEF LAT EAR.PRES (KC)
1	23.00	110.00	0.00	0.00	1.00
2	28.00	55.00	0.00	0.00	1.00
3	28.00	55.00	0.00	0.00	1.00
4	23.00	38.00	0.00	0.00	1.00
5	30.00	60.00	0.00	0.00	1.00

* STRATUM * NUMBER *	COEF LAT EAR.PRES (KT)	TERZAGHI BEAR.CAP. (NC)	TERZAGHI BEAR.CAP. (NQ)	BOTTOM ELEVATION FEET	SOIL/PILE FRIC.ANG. DEG.	SOIL TYPE
1	0.70	0.00	10.00	0.00		CH
2	0.70	0.00	10.00	-5.00		ML
3	0.70	0.00	10.00	-12.00		ML
4	0.70	0.00	10.00	-20.00		CH
5	0.70	0.00	10.00	-60.00		SP

(Continued)

STRAT CAPAC *NUM TENSION * TONS	EL.TIP (FT)	COH/ADH. RESISTAN	FRICTION COMPRESS TONS	RESISTAN TENSION TONS	END BEARING TONS	PILE CAPAC IN COMPRS. TONS	PILE IN TONS
1	5.000	0.000	0.726	0.508	0.735	1.461	0.508
1	0.000	0.000	2.903	2.032	1.470	4.373	2.032
2	0.000				1.470	4.373	2.032
2	-2.500	0.000	4.581	3.207	1.654	6.235	3.207
2	-5.000	0.000	6.531	4.572	1.837	8.369	4.572
3	-5.000				1.837	8.369	4.572
3	-8.500	0.000	9.470	6.629	1.837	11.308	6.629
3	-12.000	0.000	12.487	8.741	1.837	14.324	8.741
4	-12.000				1.837	14.324	8.741
4	-16.000	0.000	15.418	10.793	1.837	17.256	10.793
4	-20.000	0.000	18.342	12.840	1.837	20.180	12.840
5	-20.000				1.837	20.180	12.840
5	-40.000	0.000	37.315	26.121	1.837	39.153	26.121
5	-60.000	0.000	56.729	39.710	1.837	58.566	39.710

* RUN COMPLETED *   1 TON = 2000 LBS.

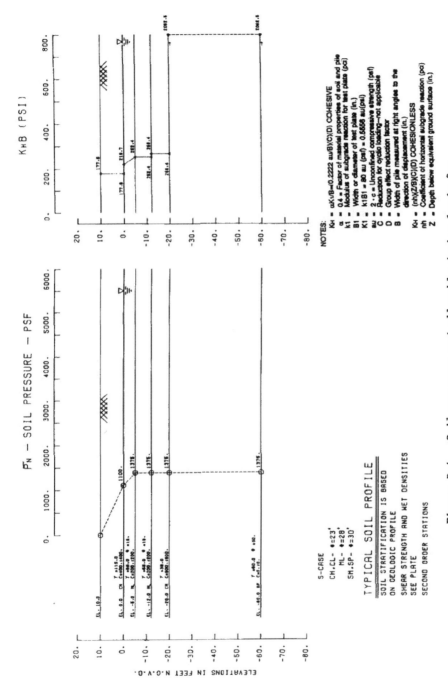

Figure D-4. Soil pressure and allowable design loads for
layered clay, silt, and sand subgrade (Continued)

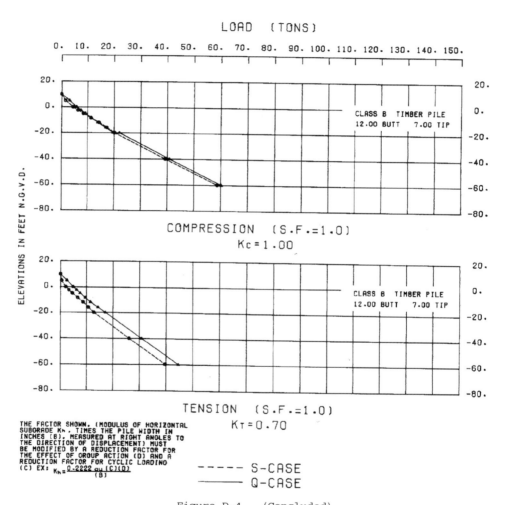

LOAD (TONS)

COMPRESSION (S.F.=1.0)
$K_C = 1.00$

TENSION (S.F.=1.0)
$K_T = 0.70$

THE FACTOR SHOWN, (MODULUS OF HORIZONTAL
SUBGRADE K_h, TIMES THE PILE WIDTH IN
INCHES (B), MEASURED AT RIGHT ANGLES TO
THE DIRECTION OF DISPLACEMENT) MUST
BE MODIFIED BY A REDUCTION FACTOR FOR
THE EFFECT OF GROUP ACTION (D) AND A
REDUCTION FACTOR FOR CYCLIC LOADING
(C) EX: $K_h = \frac{0.2222 \, qu \, (C)(D)}{(B)}$

- - - - - S-CASE
————— Q-CASE

Figure D-4. (Concluded)

Made in the USA
Las Vegas, NV
06 December 2023

82161606R00109